Andrew L. Bouwhuis
Library
Canisius College

Donated By:

Peter & Mary Lou Vogt

BRONZE SCULPTURE OF
"LES ANIMALIERS"
REFERENCE AND PRICE GUIDE

by Jane Horswell

The Antique Collectors' Club,
Clopton,
Woodbridge,
Suffolk.

© Copyright 1971
World copyright reserved

The copyright of this book is owned
jointly by the Antique Collectors' Club
and Jane Horswell

No part of this book may be reproduced
in any form without permission from the
publisher, except for the quotation of
brief passages in criticism.

While every care has been exercised in
compilation of the information contained
in this book, neither the author nor the
Antique Collectors' Club accept any
liability for loss, damage or expense
incurred by reliance placed on the
information contained in this book.

Printed in England by
Baron Publishing, Woodbridge, Suffolk.

FOREWORD
The Antique Collectors' Club

The Price Guide Series, in which this volume is the eighth, is an off-shoot of the Antique Collectors' Club which was founded in 1966. Such was, and is, the keeness of the 5,500 members of the Club to discover more about quality and value of antiques, that the production of this series was undertaken.

The Club magazine which is sent free to all members contains four main sections. These are:-

1) News of the activities of the 50 Regional Clubs where members meet to discuss and handle antiques as well as listening to lectures and making visits.

2) Illustrated sale reports showing a selection of the prices realised at auction during the preceding month. Pictures of items sold at auction are so much more illuminating than the mere recital of unillustrated catalogue descriptions and prices.

3) Several liberally illustrated articles each month by practical experts which contain valuable information on prices, price trends, features of value, investment potential, fakes, etc. in fact the information which is essential to the buyer but is not normally covered in commercial magazines on antiques.

4) A large 'For Sale' section through which members buy and sell antiques among themselves. This section has, perhaps given rise to the myth that the Club is anti-trade. It is not and this fact is clearly stated in the magazine.

It is the firm belief of all those connected with the running of the Club that the more collectors know about antiques the more enjoyment they will derive, and the more the antique trade in this country will flourish. Accordingly, from time to time the magazine carries articles on animalier bronzes by the author of the Guide and by other specialist contributors.

The annual subscription of the Antique Collectors' Club is £3.20 per annum.

For Collectors   *By Collectors*   *About Collecting*

The Antique Collectors' Club
Clopton, Woodbridge,
Suffolk.

PRICE REVISION LISTS

The prices given in this book will be outdated by the passage of time. It is suggested that, in order to keep the values up to date, readers should obtain the price revision lists as they are published. These contain a complete revaluation of all pieces shown in the book.

The first revision list will be published on 1st October, 1972, and it is proposed to publish revisions annually. The cost is £1.50 per annum by Banker's Order and the revision list is available from:

The Antique Collectors' Club

Clopton, Woodbridge,

Suffolk.

ACKNOWLEDGEMENTS

The Antique Collectors Club are doing excellent work with their interesting and informative publications on a variety of subjects and I have had considerable enjoyment compiling this book on the bronze sculpture of the French Animalier artists. It is a good subject – sculpture that can be understood has been collected through many centuries and will continue to be. I could have spent another year at least – there is so much to cover – and I ask readers to forgive any omissions.

I am indebted to Professor Glenn Benge for his interest and help over a considerable period and the use of his researches on Antoine-Louis Barye, and also to the following who have given me help or advice or both:– M. Jacques Thirion, of the Musée du Louvre, M. Le Conservateur of the Musée Carnavalet, Mr. Charles Avery of the Victoria and Albert Museum, Mr. G.L. Taylor of the Ashmolean Museum, Mr. William Johnston of the Walters Art Gallery, Mr. Richard Franke Goldman, President of the Peabody Institute, Mr. Michael Gaskin of Art Bronze Foundry, Dr. Andrew Ciechanowiecki and M. Jacques Chalom.

My thanks are also due to Mary Haynes for her help with much translation, to the efficient and interested library staff at the Musée des Arts Decoratifs, to Dorothy Parr and Betty Sobey who interpreted my notes, to Sotheby's, Christie's, Henry Spencer of Retford, Mr. and Mrs. Glenn Miller and Mr. Morton Lee for the use of their prints, and to the photographers, Alan Holmes, who worked on most of the sculpture illustrated, and Peter White.

CONTENTS

	Page
Introduction	i
Explanatory note on illustrations and prices	iv

Biographies and Illustrations:-

Antoine-Louis Barye	1
Christophe Fratin	81
Pierre Jules Mêne	103
Marie Rosalie (Rosa) Bonheur	174
Emmanuel Frémiet	181
Isidore Bonheur	201
Jules Moigniez	217
Additional Sculptors — with Illustrations	249
Additional Known Sculptors, illustrations of unsigned bronzes also Porcelain	292
Founding Practice — Additional Photographs	303
Signatures	310
Details of Other Known Founders	312
Appendix A — Methods of Casting (cire, sand, etc.,)	314
Appendix B — Lists of Sculptors' works — Salon Exhibits etc.	316
Bibliography	339

INTRODUCTION

Les Animaliers' — strictly the nineteenth century French school of sculptors — flourished from the 1830's until the turn of the century. Barye, Fratin, Mêne, Frémiet, Moigniez and the Bonheurs, the foremost amongst them, are names that are as familiar now as they were to the collectors of their own period. The exceptional success of this school in their own lifetime is rare in the history of sculpture. Artists were rarely 'men of means' and could not establish themselves without help and grand patrons or state commissions were most necessary if they were to make a living. Prior to the nineteenth century successful sculptors naturally emerged, but there must have been many with talent who spent their working lives in the studios of the celebrated. The rich and noble patrons were few and, amongst the general public, there had never been many with the necessary appreciation of art or the funds to purchase it. This was all greatly changed by the advances made in industry, with the subsequent wider distribution of wealth — with more education and with money in their pockets, a large range of new patrons emerged. As well as the private individuals there were also the many public bodies who were interested to buy or commission sculpture.

A vital point when discussing the success of the animalier school, was the talent and tenacity of Antoine-Louis Barye. He was the corner-stone, and his personal achievements had a bearing on the success and popularity of the animalier school. These were also enhanced by several coincidental factors: the revival of interest in small bronze sculptures, the technical advancement made in the early nineteenth century, the greater distribution of wealth, and a subtle desire for change, such as that which occurred after the first world war. war.

In 1831 Barye exhibited his "Tiger and Gavial". This work caused a furore. Since the ancient world, the animal had had a somewhat secondary position in sculpture; it was either symbolic, decorative, or simply used as an appendage to the more important human form, and Barye's preoccupation with it in 'pride of place' brought him much strong criticism from the academics of the art world. Knowing nothing of zoology, they accused him of 'inventing nature', nevertheless these outbursts did no harm, for the old maxim that 'any criticism is better than none at all' applied. Remarks such as this, together with a great deal of justifiable praise from the more courageous, made certain that this work was not ignored. One partial critic wrote of the "Tiger and Gavial", 'the reality of this piece is so vivid that one feels followed by the odour of the menagerie.' This model was extremely important in the history of nineteenth century French sculpture, as it not only brought recognition for Barye, but also for the subject he chose.

Indirectly the Sèvres Porcelain Company was in some way responsible for the renewal of interest in small bronzes. In the eighteenth century famous sculptors had composed groups for them on the small scale which were reproduced in 'biscuit' or *parien* porcelain. These were extremely popular, and later there was to be an interest and demand for the original models themselves, and by the early part of the nineteenth century the small bronze was once more in favour. The animaliers profited by this, and the Salon exhibitions of the 1830's onwards saw, not only work by Barye, but by other animal artists exhibiting for the first time. Fratin in 1831, Pierre Jules Mêne in 1838, Rosa Bonheur in 1841 and Frémiet in 1843, followed by many others. Their exhibits were seldom 'larger than life' — almost always expected of important sculpture before — but on the scale that we are familiar with today.

The expanding nineteenth century clientèle demanded good workmanship — they certainly had their houses and their furniture built to last — and by the end of the first half of the century there were many foundries, both large and small, in Paris alone, employing several thousand specialist craftsmen working on bronze sculpture generally for this eager public. Amongst them, the two firms formed by the brothers Susse and by Ferdinand Barbédienne in the early part of the nineteenth century benefited particularly from the evolutions in foundry practice and from the skill of the craftsmen that they were able to employ. Businesslike, also discriminating, these men assured productions of taste and quality, whether they were editing the 'antique' works of the past or those of their contemporaries, and it was natural that they were anxious to include work by the increasingly popular animaliers amongst their productions. In most cases this did not happen until after 1850. However, the important question of the methods of the founding of animalier sculpture and by whom, is covered separately, as these sculptures vary considerably in the quality of the casting, some being mediocre and some exceptional.

In the case of Barye, Mêne, Frémiet and Cain, who cast much of their own work, there are some bronzes that can be called 'autograph' — modelled, cast and finished by the artist himself. They were craftsmen-sculptors and when one looks into their backgrounds, although they had tuition in either painting or sculpture, or both, almost without exception the important artists were from families who were in some way connected with the working of metal. Barye was a goldsmith's son, Mêne a metal turner's, Moigniez a metal gilder's, Frémiet the grandson of a coppersmith and Rosa and Isidore Bonheur were relations of the founder Hippolyte Peyrol. Talent was obviously the first necessity but, after discovering this, these associations are bound to have been of considerable help. We know that Barye, Mêne (who was later joined by his son-in-law, Cain), and Frémiet, busy producing their own work in their own foundries, found it worthwhile to issue catalogues and sell their work direct to the public. As well as the desire in these new collectors to have work of the finest quality, the models themselves had a special appeal in nineteenth century France as, strange as it may seem, considering the bitter struggles that had continuously ensued with England and which culminated in Waterloo, the taste was for everything 'anglophile'. The English 'scene' is much in evidence in many models by the animalier artists, and it was not surprising that the popularity of this work, once assured in France, was to spread to England and later to America.

It may seem strange, but it is not altogether remarkable, that France produced this school of animal sculptors (there were only two Englishmen in this field and one a pupil of Frémiet) in a period when animal painting was flourishing in England. The 'change' in the world of art and letters was initiated in Europe where the 'romantic' movement was on the way, expressing itself in the writing of Madame de Stael and Victor Hugo, and in the paintings of Géricault and Delacroix. In sculpture it was Barye who first broke away from the stiff neo-classical style of Napoleon's First Empire, with its solemn dignity and adulation of the human form, and, turning to nature, brought imagination and vision to his sculpture of animals.

The Paris zoo, 'Les Jardins des Plantes' was Barye's field of study, laboratory, and inspiration, and his scrupulous observations of animal behaviour and his study of their anatomy set a pattern for the other contemporary artists, who benefited from his researches. There is little doubt that as a result of Barye's initial determination to gain recognition for the animal's place in art, a whole school flourished and from this stems the tradition of animal sculpture that has continued now for over one hundred and fifty years. Degas, Troubetskoy, Bugatti, Pompon, the American Haseltine and others can be followed through to the talented sculptors working in this field today. The sculpture produced by this school itself, the animaliers of the nineteenth century France, is now unique, and will have a place in the history of French sculpture.

EXPLANATORY NOTE ON ILLUSTRATIONS AND PRICES

All photographs, unless otherwise stated, are the copyright of the Sladmore Gallery — London.

The written matter that concerns each illustration is divided as follows:-

1) Title in English, also French if the artist's description is known.

2) Salon Exhibition and other catalogued dates where available.

3) Description of the actual bronze illustrated.

4) A brief opinion of the original sculptured model, not the illustrated example.

5) Additional information — museums etc.

6) The suggested price.

The price of animalier sculpture is based initially on who is the artist, and on the subject matter and the rarity, but a most essential element is the quality of the cast itself which makes it impossible to put an exact price on any bronze without examination. The values I have given are not for the actual bronzes illustrated, but represent the average price for a reasonably good cast to a really fine work, not anything obviously inferior.

The use of the word 'rare' means that I have seen from one to three examples only (with the possible exceptions of museums) over a period of twenty years.

Antoine-Louis BARYE 1796 - 1875

The complete history of Barye's life and his achievements in the world of sculpture is well covered by his biographers. It is too wide a subject to attempt within the confines of this book, which will deal entirely with Barye as an animal artist, with pertinent details about the man, his life and his times, which could be of interest to collectors of his work and of animalier sculpture as a whole.

Barye was born in 1796. The son of a goldsmith, he received little education to speak of, and was apprenticed to the steel engraver, Fourrier, at the age of thirteen. This involved making buttons, badges, helmets and other decorative paraphernalia for the French army, but there was nevertheless more artistic work to come. Fourrier executed commissions for the goldsmith Biennais, embossing historical scenes in bas-relief for the decoration of gold snuff-boxes destined for the Emperor Napoleon to present to other Sovereigns. In 1812 the armies were once more mobilised and all young men were needed. Barye, aged sixteen and a half, instead of carrying arms at once, found himself doing familiar work — the military 'reliefs' required for the forthcoming battles — Mont Cenis, Cherbourg and Coblenz. Later, after the capitulation of Paris and his release from the army, Barye is reputed to have told the writer Théophile Silvestre, in 1814, "I took up again my profession of chiselling, but I was continually troubled by my ambition to be a sculptor. I applied myself completely to drawing and modelling, but as I was not busy, I did not know how to find a master, nor how to live while studying". In spite of an unhopeful start he found a master, in fact, two, and although there were various surmountable setbacks in his career, he achieved his ambition and became a sculptor.

In 1816 he was accepted as a pupil by the sculptor, Bosio, and three months later by the painter, Gros. Presumably this was a part-time arrangement, as Barye is known to have frequented the studios of both these men at this time and also to have carried on with his trade as an engraver. Under Bosio, who was a sculptor of the neo-classic school of Canova and David, with the conventions, affectations and restrictions that it imposed, Barye nevertheless learnt the technique of sculpture and it is possible that his reactions against the limitations of this style gave him the urge for adventure in composition, using more natural and free forms. With the painter Gros, Barye must have been in complete accord. Gros's exciting canvases depicting tumultuous battle scenes — rearing horses with flowing manes, and riders with plumes billowing — had the romantic conception which Barye was later to capture so magnificently in sculpture. Indefatigable in his pursuit of a career, he was also attending classes at L'Ecole des Beaux Arts, and in 1819 he attempted the Prix de Rome, the scholarship inaugurated by the Medici's. He entered a medal engraving "The Milon of Croton", which, although confined to the bas-relief, has a certain interest, as Barye portrays an animal for the first time — a lion sinking its jaws into the thighs of the man. While the lion is intended to take second place and is partly obscured by the human form, it strangely holds one's interest as the dominant character. Barye failed to win the prize, but was invited to enter sculptures in the three successive years, achieving a second award, but never the elusive first. If he had gone to Rome he would have received the government grant, essential to an artist without means. The loss of this prize was a bitter disappointment, and Barye left the college. There may be the consolation that study in Italy could possibly have curbed the very forceful originality which was to come, but, be that as it may, instead he remained in Paris and became the friend of such men as Saint-Beuve, Dumas, Millet, Gérome, Géricault and Delacroix, joining their circle of artists and men of letters. Nevertheless, Barye still had his livelihood to consider, and had to be content to model small animals for Fauconnier, goldsmith to the court until 1831. Some of these unsigned works commanded certain attention and praise when exhibited at an exhibition in 1825, and it is reputed that, later, after Barye was famous and established, he was asked to sign them by Fauconnier's heirs. There is little known of them now, except that the model, "Hercules and the Wild Boar", probably dates from this early period of his work.

As a result of Barye's intellectual curiosity, the artisan was becoming the artist; he was adapting the bold and flowing lines of romantic painting to sculpture, also studying the essence of things natural and scientific. He spent any available time at the Paris zoo, 'the gardens' as it was called, where he was a well known figure. The officials there were interested and intrigued by his studies, and co-operated in every way. They provided him with carcasses for dissection when possible and sometimes with food for himself – Père Rousseau, one of the keepers, who on occasions gave Barye bread meant for his bears, claims that the two important things in his life were showing the Emperor around the zoo and the eventual success of Barye, whom he helped and encouraged.

Barye exhibited for the first time at the Salon of 1831 when still working for Fauconnier. The success of his group "The Tiger devouring a Gavial", changed everything. He had not dropped an artistic 'clanger' as hoped for by some, but created a *tour de force*. This group was purchased by the Minister of the Interior for the Luxembourg Palace, and, from this time on, receiving commissions and acquiring patrons, Barye's stature as a sculptor increased. In 1832 he executed another sensational work, the "Lion and Serpent", in plaster, which was shown at the Salon of 1833. This exhibition was to be a great step forward for Barye. "The Lion" was acquired by the State*; also, King Louis-Phillipe's sons, the Duc d'Orléans and the Duc de Nemours, purchased several works. After the exhibition Barye received the Croix de la Légion d'Honneur by royal order, on the 30th April.

In 1834, Barye and the eminent designer, M. Aimée Chenvenard, received an important commission from the Duc d'Orléans which is interesting because of its subject, animals, and also because of its rather bizarre history. The press announced it as a large centrepiece for the Duke's dining table, to consist of about fifteen major groups of figures and animals to be founded in gold and silver by the 'lost wax' method. The smooth surfaces (Barye's bronzes?) were to be enamelled in the Florentine style and encrusted with many fine stones of brilliant colours – the mosaic base, formed of malachite, lapis-lazuli and other precious marbles, was to be twenty-one feet long and five feet wide. When the first piece was ready, presumably the base only, it required a special machine to take it to the Tuileries. When placed in position the table collapsed under its weight, and the Duke sanctioned the construction of a larger one that would bear it. This failed to solve the problem, as it was found that there was then no room left for chairs. M. Chenvenard requested the enlargement

*A life-size bronze, cast by the Gonons, it is now in the Salle de Barye – Louvre.

of the dining room, but this was refused. The Duke must have made his decision by that time that he wanted something more simple for everyday use, and the project was abandoned.

By 1837 Barye had completed five groups of 'hunts' originally designed for this centre piece, and these were cast in bronze by Honoré Gonon. Barye sent them to the Salon of that year, where the panel of judges refused them. They were the Duke's property, and, in spite of his involvement, and the pressure he put on his father, the King, to intervene, they were not accepted – it is said more for political than artistic reasons. Barye was indignant and refused to exhibit at the Salon for over ten years, in silent protest against this decision. (He was to be elected to the panel himself some eighteen years later.)

A monumental project which was underfoot in the late 1830's, and which gives another example of Barye's high principles and strength of character, was the crowning of the Arc de Triomphe at the Etoile. The general opinion was that a vast eagle with wings outspread, to be placed at the highest point, would be appropriate. Barye was consulted and made his suggestions, to which the King agreed. There would be the eagle, also other statuary subservient to it, depicting the various conquests of the 'glorious empire', which naturally included horses and riders. When Barye heard that Rude would do the riders that would sit on his horses, he refused to take part. Barye was not without important work nevertheless, as, prior to this, he had been commissioned to do the famous "Striding Lion" and other animals to decorate the July Column in the Place de la Bastille, also his "Seated Lion", which he had shown in the Salon of 1836, in plaster, was commissioned in bronze by the State for the Tuileries Gardens. This life-size work was not completed until 1846 – it was a *cire perdue* cast by the Gonons, who were close associates of Barye in the early part of his career.

In 1839 Barye borrowed money and set up his own foundry and sales gallery. He was in a reasonably firm position by then with commissions from the State and from his royal patrons, executing the "Roger et Angélique" group with its two adjacent candelabra for the Duc de Montpensier, as well as completing the "Seated Lion" for the Tuileries. It is difficult to determine exactly which of his small sculptures were created in the 40's, as he had declined to send exhibits to the Salon, but by now he had started producing animal groups for the general public. These sculptures, some modelled initially in the thirties, were cast on a comparatively large scale during this period, resulting in the publication of his first catalogue from No. 6 Rue de Boulogne on the 1st September, 1847. He had certainly been busy, as some 107 different works were listed, in four sections, in a price range of 5 francs for a rabbit to over 1,000 francs for the "Roger et Angélique" group. In some cases there was only one cast available, in other cases more, and all casts of each group listed in this catalogue were numbered as they left the foundry.

From 1848, Barye held a post as curator at the Louvre, which he was forced to relinquish in 1850 due to the political turmoil of the time. He left in some haste, bundling everything from the workshop he had there into a handcart. A plaster model was reputed to have been smashed during this episode: it was to have been a companion piece to the large "Theseus Fighting the Centaur Bienor" which had been bought by the Ministry of the Interior in 1849. There were obviously no half measures with Barye either in his involvement with political matters or in his attacks upon the 'powers that be' at the Salon. A protest against this body, which he organised and signed earlier, along with other 'romantics' including Daumier, Corbet,

Rousseau and Decamps, must have had some effect, as he exhibited again in 1850, showing his "Jaguar devouring a Hare" — an outstanding work and one that ranks high among his many inspired and successful compositions.

At this time the rather puzzling associations with M. Emile Martin began. It is said that Barye owed him money, and that consequently Martin took over the handling of his affairs, but there is considerable doubt about this. Barye was an artist, not a business man, and it is possible that there was an arrangement beneficial to both sides. Certainly Martin became Barye's agent, keeping the accounts and handling the transactions, not only between customers, but between the individual founders, chisellers and other craftsmen, also the commercial firm of Barbédienne, who were involved in varying degrees in Barye's work. During this successful association with Martin, which lasted until 1858, four more catalogues were published from different addresses. Meanwhile Barye, apart from important commissions, accepted teaching appointments during this period. He was appointed Professor of Drawing for Zoology at the Museum of Natural History in 1854, a post which he held until his death, when it passed to Emmanuel Frémiet.

One of the most interesting things that happened when reviewing Barye's work today, was the arrival in Paris of the American, George A. Lucas, in 1857, followed by William T. Walters of Baltimore in 1859. Walters was just visiting, but Lucas made Paris his home for fifty years, and he and his close friend Walters became admirers and patrons of Barye, also personal friends. Both of them, authorities on art, introduced others to Barye's work. They were in close touch with such prominent Americans as Samuel Avery, J.T. Johnson, Frank Frick and Cyrus J. Lawrence, who were either important private collectors or became benefactors of the Metropolitan Museum. William Corcoran, a banker and great art collector, had by that time established the Corcoran Art Gallery in Washington and William Walters commissioned from Barye, on Corcoran's behalf, a cast of all his works, of which about 120 were achieved. Lucas's collection is now in the Baltimore Museum of Art (on loan from the Maryland Institute), and the fine collection of William Walters, which was later enlarged by his son, is in the Walters Art Gallery, also in Baltimore. Walters was not only a great patron of the arts, but also an 'animal' man. He introduced the Percheron horse to America, and this may have had some bearing on his admiration for Barye and his work. After Barye's death in 1875, these American friends and patrons did much to honour his memory, organising exhibitions and erecting monuments to the artist, in Paris and in America.

Today, Barye's sculpture is as popular with American collectors as it was in the nineteenth century, and in England now the work of this great artist is at last appreciated. In Barye's lifetime his vivid compositions of animals in conflict, the subject of so many of his finest groups, disturbed the English, and they would look no deeper. In consequence the collections in our premier museums cannot compete with those of America and France. Maybe we can rectify this one day, as much that is 'the best of Barye' has now found its way to this country.

Barye's great original talent as an animal sculptor is without parallel. He not only astonished and disturbed the conventional attitude towards the animal world in art, but was able to give us creatures, within the confines of small sculpture, whose natural grandeur, power and dignity were enhanced.

African Elephant Charging   **Barye**
Eléphant du Sénégal Chargeant

Listed Barye catalogue of 1855.

Bronze illustrated: Medium dark patination. Fair quality cast. Height 5 inches.

This model 'speaks for itself' — it was one of the most popular creations by the artist. It would appear that there were a fair number cast of this group, varying considerably in quality. Those by F. Barbédienne are usually good.

Examples of this work in the possession of:-

        Louvre Museum — Paris (Barye Exhibition 1956/7).
        Victoria & Albert Museum — London.
        Museum of Fine Arts — Boston.
        Brooklyn Museum of Art.
        Fogg Art Museum — Cambridge, U.S.A.
        Metropolitan Museum of Art — New York.
        City Art Museum — St. Louis.
        Worcester Art Museum — U.S.A.
        Corcoran Gallery — Washington.
        Walters Art Gallery — Baltimore.

Illustrated — Ballu.

Value: £250–£400

Elephant Crushing a Tiger　　　　　　　　　　　　　　　　　　　　　　　　　　**Barye**
Eléphant Ecrasant un Tigre

Model listed in Barye's second (undated) catalogue – prior to 1855.

Bronze illustrated: Pale green-brown patination – good cast. Height 8½ inches.

A powerful and imaginative group. Closely allied to the "Tiger Hunt" commissioned by the Duc d'Orléans, the model probably dates from the late 1830's.

Examples of this work in the possession of:-

    Walters Art Gallery – Baltimore.
    Boston Museum of Fine Arts.
    Brooklyn Museum of Art.

      Value: Rare £550–£950

Indian Elephant *Barye*
Elephant d'Asie

Model exhibited Salon of 1833

Bronze illustrated: Medium dark patination – very fine early cast. Height 6 inches.

Excellent model – character of the Indian, as opposed to the African, elephant well defined.

Examples of this work in the possession of:-

    Louvre Museum   Paris (Barye Exhibition 1956/7).
    Baltimore Museum of Art, Lucas Collection.
    Museum of Fine Arts – Boston.
    Brooklyn Museum of Art.
    Walters Art Gallery – Baltimore.

        Value: Rare £250–£400

Eagle on a Rock with a Dead Heron Barye
Aigle prés d'un Heron Mort

Model listed in Barye's catalogue of 1855.

Bronze illustrated: Rich dark patination – green-brown undertones. Very fine cast. Signed A – L BARYE. Height 12 inches approx. Exhibited Barye Exposition 1900. Ex. Collection George Lutz.

A highly successful composition – the model probably dates from about 1836 (the L'Etoile project).

Examples of this work in the possession of:-

        Baltimore Museum of Art, Lucas Collection.
        Metropolitan Museum of Art – New York.
        Philadelphia Museum of Art.
        Petit Palais, Paris. (Bronze patinated plaster exhibited Barye exhibition Louvre, Paris, 1956/7).
        Victoria & Albert Museum – London.
        Corcoran Gallery – Washington.
        Walters Art Gallery – Baltimore.

Illustrated – Ballu

Value: Rare £700–£1,000

Stag and Two Hounds　　　　　　　　　　　　　　　　　　　　　　　　　　　**Barye**

The property of the Ashmolean Museum, Oxford, this bronze bears the signature and date – BARYE 1832 – also the inscription "Fondu d'un seul jet sans ciselure par Honoré Gonon et ses deux fils".

Rich brown patination – The surface detail of the original model itself is well-defined and the 'waxy' look associated with 'cire perdue' casting can be seen clearly in the illustration of this very fine work. It is quite probably a *pièce unique*.

Illustration – courtesy of the Museum.

Ten-Pointed Stag Brought Down by Two Scotch Hounds Barye
Un Cerf Dix-cors Terrassé par Deux Lévriers d'Ecosse

Listed in Barye's first catalogue of 1847.

Three bronze illustrated: First two exceptional examples are numbered casts from the first half of the nineteenth century. Close examination will reveal considerable differences – treatment of base foliage, the base itself, chiselled detail etc. The patination is almost black on one and a pale bronze on the other. Equally fine is the third group illustrated (by courtesy of Christie's). It shows a very marked variation in the position of the stag and the topmost hound and again a different base and detail.

These are very fine early models by Barye, although the third group is altogether more active; in the first two the exhaustion and defeat of the stag is beautifully handled. Barye exhibited a model "Stag Brought Down by Two Large Hounds (Lévriers)" at the Salon of 1833 which may be the bronze in the Ashmolean, Oxford, (previous group), as details, apart from the title, are not available. The other groups could be earlier than 1832 – the conception is more realistic and less fluid than the Ashmolean bronze. These examples give an idea of the considerable differences that can appear on each individual cast. Length approx. 21¼ inches

Examples of these works are in the possession of:-

Metropolitan Museum of Art – New York.
Corcoran Gallery – Washington.
Walters Art Gallery – Baltimore.

Value: These groups are rare: if found in this quality, £750–£1000.

Fawn Recumbent
Faon Couchée
                                                Barye

Listed in Barye's first catalogue of 1847.

Bronze illustrated: Rich dark brown/green patination. Good cast – probably 'lost wax'. Height 2¾ inches.

Examples of this work are in the possession of:-

    Brooklyn Museum of Art.
    Philadelphia Museum of Art.

                Value: Rare £150–£260

Stag Seized by Wolf *Barye*
Loup Tenant un Cerf à la Gorge

Model listed in Barye's first catalogue of 1847.

Bronze illustrated: Rich brown patination — Exceptionally fine and detailed chiselling. Height 8 inches.

This model is a beautifully detailed work. The positioning of the wolf shows its character admirably.

Examples of this work in the possession of:-
      Louvre Museum — Paris (Barye Exhibition 1956/7.
        Dated 1843. Plaster and wax).
      Baltimore Museum of Art, Lucas Collection.
      Brooklyn Museum of Art.
      Corcoran Gallery — Washington.

          Value: Rare £1,000–£1,500

Elk Attacked by a Lynx   **Barye**
Elan Surpris par un Lynx

Plaster study exhibited Salon of 1834.
Listed in Barye's first catalogue of 1847.

Bronze illustrated: Dark patination – An exceptional cast by the 'lost wax' method. Height 8½ inches.

A very dramatic composition – great contrast between the large elk and its small enemy.

Examples of this work in the possession of:--

       Baltimore Museum of Art, Lucas Collection.
       Corcoran Gallery – Washington.
       Walters Art Gallery – Baltimore.

Illustrated – Ballu.

                 Value: Rare – £1,100–£1,650.

Kevel Antelope   Barye

Listed in Barye's catalogue of 1855.

Bronze illustrated: Rich brown black patination – Fine quality bronze.

Pleasing model – Probably dates from the late 1830's.

Example of this work in the possession of:-
       Philadelphia Museum of Art.

              Value: Rare   £250–£350

Gazelle **Barye**
Gazelle d'Ethiopie

Model dates from 1837 – Listed as "Gazelle" in Barye's catalogue of 1847 (full title 1855).

Bronze illustrated: Medium dark patination. A dated (1837) cast. These dated works are invariably of the finest quality. Height 3½ inches.

A particularly attractive model among the many small sculptures executed by Barye in this period.

Examples of this work in the possession of:-

       Louvre Museum – Paris (Barye Exhibition 1956/7).
       Walters Art Gallery – Baltimore.

           Value: Rare   £350–£500

Stag, Left Foreleg Raised *Barye*
Un Cerf, la Jambe Levée

Model listed in Barye's first catalogue of 1847.

Bronze illustrated: Dark patination – Exceptionally fine cast of the 1830's. Stamped Barye 22. Height 7½ inches.

A proud stag – one of Barye's best models of this subject. Dates from about 1838.

Examples of this work in the possession of:-

    University of Michigan.
    Brooklyn Museum of Art.
    Fogg Art Museum – Cambridge, U.S.A.
    Philadelphia Museum of Art.
    Corcoran Gallery – Washington.
    Walters Art Gallery – Baltimore.

                Value: £300–£550

Stag Walking – Right Foreleg Raised Barye
Un Cerf qui Marche

Model listed in Barye's first catalogue of 1847.

Bronze illustrated: Medium dark patination – very plain, late cast – antlers have suffered considerable damage. Height 5½ inches approx.

Rather a dull model – the stag has a glazed look which would be improved on a better cast.

Examples of this work in the possession of:-

        Fogg Art Museum – Cambridge, U.S.A.
        Philadelphia Museum of Art.
        Walters Art Gallery – Baltimore.

            Value: £150–£250

Stag Walking – Right Rear Foot Raised Barye
Cerf Axis

Model listed in Barye's first catalogue of 1847.

Bronze illustrated: Medium dark patination – nice quality cast founded by F. Barbédienne. Height 6½ inches.

One of Barye's earlier models, about 1820–1830. (The deformed antlers of this deer suggest that the actual subject was in captivity).

Example of this work in the possession of:-

    Philadelphia Museum of Art.

Value: Rare £200–£350

Wolf Caught in a Trap   Barye
Loup Pris au Piège

Model listed in Barye's catalogue of 1855 (new work).

Bronze illustrated: Mid bronze patination – Very fine detailed cast. Height 4½ inches.

This work, although listed as a new addition in 1855, was probably originally modelled in the 1830's.

Examples of this work in the possession of:-

        The Louvre Museum – Paris.
        Brooklyn Museum of Art, Lucas Collection.
        Fogg Art Museum – Cambridge, U.S.A.
        Corcoran Gallery – Washington.

                Value: £150–£250

Turtle
Tortue

Barye

Circa 1820's
Listed (No. 1) in Barye's first catalogue of 1847.

Bronze illustrated: Light bronze patination. Very good quality early cast. Stamped Barye – length 4 inches approx.

This model, with or without a base, probably dates from the 1820's.

Examples of this work in the possession of:-

  The Louvre Museum – Paris.
  Philadelphia Museum of Art.
  Corcoran Gallery – Washington.
  Walters Art Gallery – Baltimore.
    Value: £60–£150

Bear's Head

Bronze illustrated: Brown patination. Good cast. Height 3 inches.

Extremely unusual and rare model. Probably slightly later execution than the 'bears' of the 1830's.

 Value: Rare £50–£100

Bear Lying on its Back
Ours Couché sur le Dos

Barye

Model listed in Barye's catalogue of 1847.

Bronze illustrated: Medium dark patination – brass evident. Exceptional detail – a very fine early cast. Height 3 inches.

This model and other groups of bears date from the 1830's.

Examples of this work in the possession of:-

        Louvre Museum – Paris (Barye Exhibition 1956/7).
        Victoria & Albert Museum – London.
        Baltimore Museum of Art, Lucas Collection.
        Brooklyn Museum of Art.
        Worcester Art Museum.
        Corcoran Gallery – Washington.
        Walters Art Gallery – Baltimore.

          Value: £100–£300 Late re-casts appear on the market.

Dromedary of Egypt – Harnessed Barye
Dromadaire Harnaché d'Egypt

Model listed in Barye's catalogue of 1855 (new work).

Bronze illustrated: Pale bronze patination. Rather a 'plain' cast – detail of harness could be crisper altogether. Height 6½ inches.

There is a variation of this model with an Arab seated in the saddle.

Examples of this work in the possession of:-

      Louvre Museum – Paris (Barye Exhibition 1956/57).
      Baltimore Museum of Art, Lucas Collection.
      Brooklyn Museum of Art.
      Fogg Art Museum, Cambridge, U.S.A.
      Walters Art Gallery – Baltimore.

Illustrated – Ballu

                   Value: £200–£300

Dromedary of Egypt *Barye*
Dromadaire debout

Model listed in Barye's catalogue of 1855.

Bronze illustrated: Rich dark patination showing plenty of depth. Very fine cast indeed. Height 8 inches approx.

Model, probably executed between 1847 and 1855 as not listed in Barye's first catalogue.

Examples of this work in the possession of:-

        Petit Palais – Paris (Plaster Barye Exhibition Louvre 1956/57).
        Baltimore Museum of Art, Lucas Collection.
        Brooklyn Museum of Art.
        Detroit Institute of Arts.
        Los Angeles County Museum of Art.
        Corcoran Gallery – Washington.
        Walters Art Gallery – Baltimore.

                Value: Rare £350–£500

Standing Basset
Un Chien Basset Debout

Barye

Model circa 1830's
Listed in Barye's first catalogue of 1847.

Bronze illustrated: Medium dark patination. Exceptionally fine quality cast. Height 5 inches approx.

Example of this work in the possession of:-

       Baltimore Museum of Art, Lucas Collection.
       Brooklyn Museum of Art.
       Corcoran Gallery – Washington.
       Walters Art Gallery – Baltimore.

Illustrated – Ballu.

Value: £250–£400

Seated Basset **Barye**
Un Chien Basset Assis

Model circa 1830's
Listed in Barye's first catalogue of 1st September, 1847.

Bronze illustrated: Rich dark green patination. Not a very fine cast. F. Barbédienne — founder.

A popular and commercial model by Barye.

Example of this work in the possession of:-

        The Louvre Museum — Paris (Barye exhibition 1956/57).
        Brooklyn Museum of Art.
        Baltimore Museum of Art, Lucas Collection.
        Detroit Institute of Arts.
        Worcester Art Museum.
        William Rockhill Nelson Gallery, Kansas City.

Illustrated — Ballu

Value: £250–£400

Pointer
Braque

Barye

Similar model "Chien en Arrêt" Circa 1820's
Listed in Barye's catalogue of 1855.

Bronze illustrated: Rich bronze patination. Good cast. Height 3½ inches.

A variation of the earlier model "Dog on Point" which shows a rabbit concealed in the right foreground.

Example of this work in the possession of:-

        Walters Art Gallery – Baltimore.

Illustrated – Ballu

        Value: Rare £200–£300 bad re-casts are known

Greyhound Devouring Hare **Barye**
Levrette Debout Ramassant un Lièvre

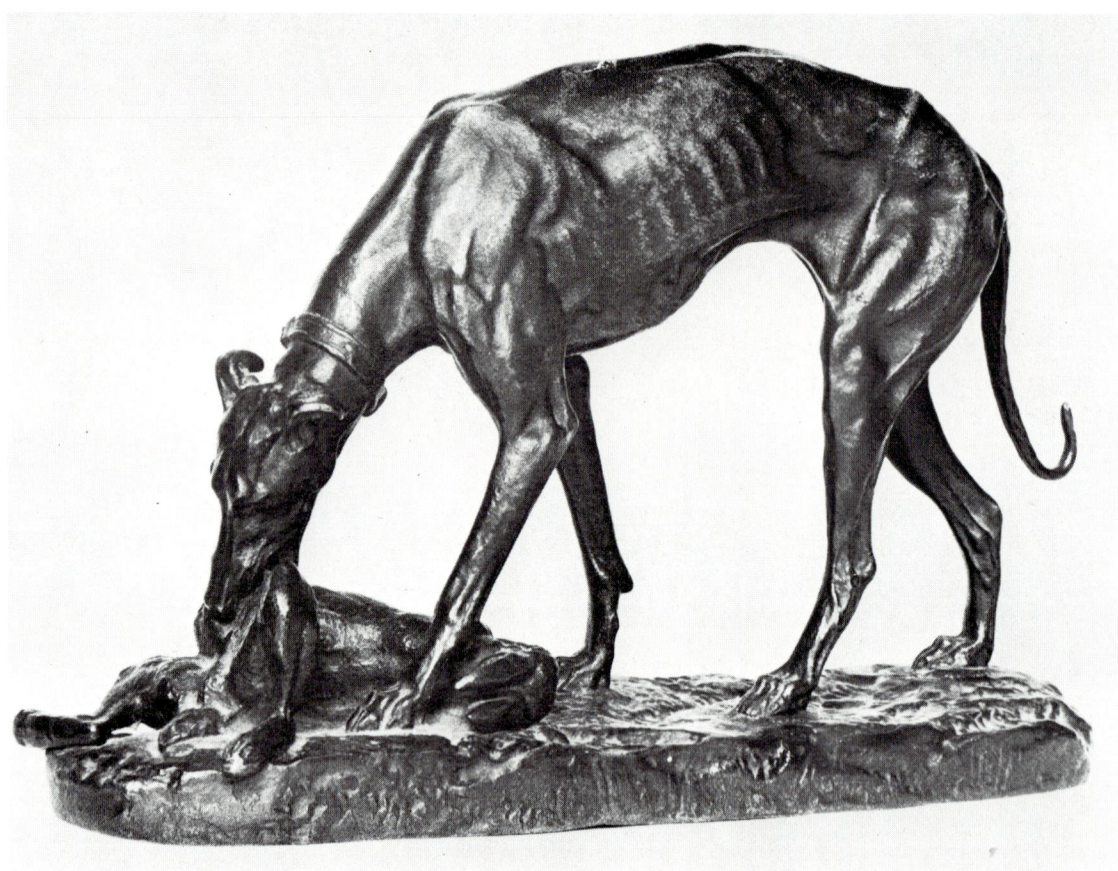

Model·circa 1830's
Listed in Barye's first catalogue of 1st September, 1847.

Bronze illustrated: Green-black patination. Rather poor surface detail. Height 8 inches.

Excellent model of the hound, but rather a dull composition compared to similar subjects by this artist.

Examples of this work in the possession of:-

       Louvre Museum. Paris (Barye exhibition 1956/57).
       Baltimore Museum of Art, Lucas Collection.
       Corcoran Gallery of Art – Washington.
       Walters Art Gallery – Baltimore.
       Victoria & Albert Museum – London.

                Value: £350–£500

Seated Hound   **Barye**

Bronze illustrated: Medium dark patination. Believed cast by the goldsmith, Levêque. An exceptional cast with good surface detail. Height 10½ inches.

A little-seen model (perhaps another version of "Tom", who was Barye's personal pet).

Value: Rare £400–£600

Rabbit — Ears Raised   Barye
Lapin, les Orielles Levées

Model listed in Barye's first catalogue of 1847.

Bronze illustrated: Medium dark patination — Not the best of casts — lacks surface detail. Height 2 inches.

Early model by Barye (circa 1820's) more unusual, but perhaps not as attractive as its counterpart, with ears lowered.

Example of this work in the possession of:-

    Walters Art Gallery — Baltimore

Illustrated — Ballu

Value: £50–£95

Rabbit — Ears Lowered
Lapin — les Oreilles Couchées

Model listed in Barye's first catalogue of 1847.

Bronze illustrated: Rich bronze and green tones giving depth of patina. Good surface detail. Height 1¾ inches.

Very successful and popular model — much sought after.

Examples of this work in the possession of:—

    The Louvre Museum — Paris.
    Philadelphia Museum of Art.
    Walters Art Gallery — Baltimore.

Illustrated — Ballu

Value: £70–£130

Seated Hare **Barye**
Un Lièvre Assis

Model circa 1820's
Listed in Barye's first catalogue of 1847.

Bronze illustrated: Dark patination. Fair quality cast. Height 4 inches.

Very popular and lively model.

Examples of this work in the possession of:—

    The Louvre Museum — Paris (Barye exhibition 1956/57).
    Brooklyn Museum of Art.
    Fogg Art Museum — Cambridge, U.S.A.
    Joslyn Art Museum — Omaha.
    Philadelphia Museum of Art.
    Walters Art Gallery — Baltimore.

    Value: £50—£100

Ram Grazing **Barye**

Model circa 1820's

Bronze illustrated: Pale yellowish patination. De Braux — founder. Height 2¾ inches.

Model probably dates from Barye's period with Fauconnier — it does not appear in his own catalogues.

Examples of this work in the possession of:—

    Joslyn Art Museum — Omaha.
    Philadelphia Museum of Art.
    Walters Art Gallery — Baltimore.

    Value: Rare £50—£80

Wounded Pheasant *Barye*
Faisan Blessé

Circa 1845
Model listed in Barye's catalogue of 1855.

Bronze illustrated: Medium dark patination with much variation in tone — Excellent crisp surface detail. Height 4½ inches.

A very fine model. The squawking, dying pheasant is finely modelled.

Example of this work in possession of:-

    Walters Art Gallery — Baltimore.

        Value: £150—£300

The Elk Hunt   **Barye**

One of the five 'Hunts' completed for the Duc d'Orléans, originally commissioned for the "Surtout de Table".

Bronze illustrated: Signed and dated Barye 1838. This bronze is the possession of the Walters Art Gallery, Baltimore. Sold at the Count Demidoff, San Donato Sale, Paris in 1870, it had been purchased by Demidoff at the 1853 Sale by the Duchesse d'Orléans. No other cast known to exist. Height 20½ inches.

Illustration — courtsey Wlaters Art Gallery, Baltimore.

The Lion Hunt — Barye

One of the five 'Hunts' completed for the Duc d'Orléans, originally commissioned for the "Surtout de Table".

Bronze illustrated: Signed and dated Barye 1837. Inscribed "Bronze d'un seul jet sans ciselure coule a l'hotel d'Angivilliers par Honoré Gonon et ses deux fils". This bronze is the possession of the Walters Art Gallery, Baltimore. As no other bronze cast is known to exist, it is probably the fifth 'hunt' bought by M. Montessier at the Duchesse d'Orléans' sale in 1853. (Three by Count Demidoff, one by M. Lutteroth, one by M. Montessier. – Lami). Height 18¼ inches.

Author's Note:

A plaster, retouched with wax entitled "La Chasse au Lion", height 50 centimetres, length 60 centimetres, is in the Louvre. It was exhibit No. 5 at the Barye Exhibition 1956–1957, and catalogued with these additional details:-

"Original model, representing Arab Horseman Defending a Buffalo wounded by Lions for the "Surtout de Table" commissioned by the Duc d'Orléans in 1834. This centrepiece, unfinished, was to have comprised of nine groups. Honoré Gonon founded five in lost wax in 1836. Refused at the Salon of 1837. Acquired at Barye's Sale by M. Jacquemart. Presented to the Museum by Zoubaloff in 1913. Exhibited at the Barye Exposition of 1875 (No. 149)."

Illustration – courtesy Walters Art Gallery, Baltimore.

The Bull Hunt — Barye

One of the five 'Hunts' completed for the Duc d'Orléans, originally commissioned for the "Surtout de Table".

Bronze illustrated: Signed and dated Barye 1838. Inscribed "Bronze d'un seul jet sans ciselure coule a l'hotel d'Angivilliers par Honoré Gonon et ses deux fils 1838". Formerly in the collection of the Duchesse d'Orléans, this bronze is the possession of the Walters Art Gallery, Baltimore. (It was bought at the Duchesse d'Orléans's sale in 1853 by M. Lutteroth. — Lami). No other bronze cast known to exist. Height 18½ inches.

Author's Note:
A plaster touched with wax entitled "La Chasse au Taureau Sauvage", height 50 centimetres, length 60 centimetres, is in the Louvre. It was exhibit No.6 at the Barye Exhibition of 1956–1957 and catalogued with these additional details:

"Original model, representing three Spanish Knights in sixteenth century costume hunting a wild bull with mastiffs ("dogues de grand race"). One of the centrepiece groups of the Duc d'Orléans. It was acquired at the Barye Sale by Jacquemart the sculptor. Presented to the Museum by Zoubaloff 1921. Exhibited at Barye Exposition of 1875 (No.150)".

Illustration — courtesy Walters Art Gallery, Baltimore.

The Tiger Hunt   **Barye**

One of the five 'Hunts' completed for the Duc d'Orléans, originally commissioned for the "Surtout de Table".

Bronze illustrated: Signed and dated Barye 1836. Inscribed "Bronze d'un jet sans ciselure fondu à l'hotel d'Angivilliers par Honoré Gonon et ses deux fils". This bronze is the possession of the Walters Art Gallery, Baltimore. Sold at the Count Demidoff Sale, Paris 1870, it had been purchased by the Count at the Duchesse d'Orléans sale in 1853. No other cast known to exist. Height 27¾ inches.

Author's Note:

A plaster and wax, height 69 centimetres entitled "Eléphant Monté par un Indien" is in the Louvre. Exhibited in the Barye Exposition 1956–1957, it is now on view in the Salle Barye. This is not the complete group – there are no tigers, only the elephant and a single figure (nearest the head) are present. Apart from a circingle there is no harness.

Illustration – courtesy Walters Art Gallery, Baltimore.

The Bear Hunt   Barye

One of the five 'Hunts' completed for the Duc d'Orléans, originally commissioned for the "Surtout de Table".

Bronze illustrated: Signed and dated Barye 1838. Marked on base "Fondu à l'hotel d'Angivilliers par Honoré Gonon et ses deux fils". This bronze is the possession of the Walters Art Gallery, Baltimore. Sold at the Count Demidoff, San Donato Sale, Paris, 1870, it had been purchased by the Count at the Duchesse d'Orléans' sale in 1853. Gonon does not state that this is in ' one pour ', but apart from the Brooklyn Museum exhibit, no other cast is known. Height 18¾ inches.

Example of this work in the possession of:-

    Walters Art Gallery — Baltimore
    Brooklyn Museum of Art.

Illustration — courtesy Walters Art Gallery, Baltimore.

Turkish Horse (a)   **Barye**
Cheval Turc

Model Circa 1838

Listed as one of a pair in Barye's first catalogue of 1847 — the pair is the same model but with the opposite foreleg raised.

Bronze illustrated: *a) Silver, over bronze, this patination is seldom seen on Barye's work. The pair to this is in the Bonnat Museum, Bayonne. They were probably a special commission for the painter Leon Bonnat, who was a friend and admirer of Barye's and formed a collection of his work. A very fine cast — Height 11¼ inches.*
b) Dark green with subtle undertones of brown, this is a superb example of Barye's patinations — Another exceptional cast. Height 11¼ inches.

This is one of Barye's finest sculptures. Romantic in conception, it shows the spirit of the paintings of Gericault and Delacroix, and became the 'prototype' of many of Barye's models of horses. The pose can be criticised as being unnatural or even impossible for this

Turkish Horse (b) *Cheval Turc* — Barye

animal but, be that as it may, it is a beautiful and exceptional work.

Example of this work in possession of:-

      Baltimore Museum of Art, Lucas Collection.
      Brooklyn Museum of Art.
      Fogg Art Museum — Cambridge, U.S.A.
      Metropolitan Museum of Art — New York.
      Wadsworth Atheneum, Hartford, Connecticut.
      Walters Art Gallery — Baltimore.
      Corcoran Gallery of Art — Washington.

Illustrated — Ballu

Value: Rare, £2,000–£3,500

Note: Ferdinand Barbédienne is known to have cast the "Turc", but probably after Barye's death, as there is considerable alteration to the model, a very unusual practice with this founder. It is altogether more stylised, similar to the later work of the American sculptor Haseltine. There is the minimum of mane compared to the detailed treatment on Barye's model, and the tail is raised more and is 'squared off'. The treatment of the body is much smoother without the muscle detail — the horse has the appearance of a heavier sort altogether, and the model is slightly larger than Barye's (about 15 inches).

Two Arab Horsemen Killing a Lion   **Barye**
Deux Cavaliers Arabes Tuant un Lion

Model circa 1830's
Listed in Barye's first catalogue of 1847/7

Bronze illustrated: Medium dark patination. Stamped BARYE 1. This is a very fine early cast. Height 14½ inches.

This model is closely allied to the "Lion Hunt" commissioned by the Duc d'Orléans. The subject was a popular one with painters of the period.

Examples of this work in the possession of:-

    Brooklyn Museum of Art.
    Metropolitan Museum of Art – New York.
    Corcoran Gallery – Washington.
    Walters Art Gallery – Baltimore.

                Value: Rare £800–£1,200

Tiger Attacking a Horse **Barye**
Cheval Attaqué par un Tigre

Model, Salon of 1835
Listed in Barye's catalogue of 1847.

Bronze illustrated: Rich dark green patination. Good detailed cast. Height 10 inches.

A very dramatic composition. The horse, a variation of the "Turc", is allied to the fallen horse of the previous group.

Example of this work in the possession of:-

        Walters Art Gallery — Baltimore.

*   Value: Rare £600–£1,000*

Horse Surprised by a Young Lion **Barye**
Un Cheval Surpris par un Lion

Model, Salon 1833
Listed in Barye's catalogue of 1847

Medium green-brown patination. Height 15 inches.

A well composed model closely allied in theme to "Tiger Attacking a Horse" – in the modelling the horse is similar to the rearing one of the group of "Arab Horsemen".

Example of this work in the possession of:-

       Louvre Museum (Barye exhibition 1956 – 1957).
       Metropolitan Museum of Art – New York.
       Walters Art Gallery – Baltimore.

Value: Rare £850–£1,200

Roger Abducting Angelica on the Hippogriff Barye
Angelique et Roger montés sur l'Hippogriffe

Listed Barye's Catalogue of 1847

Bronze illustrated: *Mid-green patination. Very fine cast. Reputed formerly the property, and sold by his heirs, of the Duc de Montpensier (with the pair of adjacent candelabras, one illustrated) for whom it was commissioned. stamped BARYE 2, 25 inches length. Gilt 'collection' numeral applied on the base (not shown).*

This important model, although not technically an animalier sculpture, is an essential part of Barye's *oeuvre*. Inspired by *Roland Furieux* by Ariosto, and the Italian romance of "Orlando the Furious", the subject matter was possibly a specific part of the duke's commission. This medieval winged creature, with the beak and talons of a hawk, is in superb contrast to the delicacy of the figures.

This is rare – the Louvre possess an example, cast by Barbédienne. The Sotheby group (with the candelabra) went to America. The Walters Art Gallery, Baltimore, also possesses an example.

Value: *The actual Sotheby group (plus candelabra) was sold for £1,000 and is a rarity. Not a typical example of Barye's work, a cast could fetch much more (or less). Rare.*

Note: *Roger, after visiting England and attending the review of Charlemagne's forces, passed Ebuda, and saw Angelica bound to a rock and about to be devoured by a sea-monster. He reached her and carried her away on his winged-griffon horse.*

Illustration – courtesy Sotheby & Co.

Candelabra of Nine Lights   **Barye**

Commissioned as part of a centrepiece by the Duc de Montpensier.
Circa 1840's

Bronze illustrated: One of a pair stamped Barye 1. These were sold by Sotheby & Co., (Lot 57 on the 13/2/60, £1,000 including the "Roger et Angélique" sculpture).

The graces adorning the candelabra were re-modelled by Barye as separate sculptures — "Minerva", "Jeune Femme Nue" and the "Jeune Femme, Les Bras Levés".

Not animalier work, they are included for their interest to collectors of Barye's work.

Illustration – courtesy Sotheby & Co.

Tartar Warrior Reigning in his Horse Barye
Guerrier Tartare Arrêtant son Cheval

Listed in Barye's catalogue of 1855

Bronze illustrated: Dark green patination. Exceptional cast by F. Barbédienne. Height 13½ inches.

Barye's "Turc" is in evidence again in this powerful and dynamic model.

Example of this work in possession of:-

        Baltimore Museum of Art, (Lucas Collection).
        Brooklyn Museum of Art.
        Walters Art Gallery, Baltimore.
        Corcoran Art Gallery, Washington.

Illustrated – Ballu

Value: Rare £800–£1,000

"The Young Bonaparte"
General Bonaparte

Barye

"The Young Bonaparte" and the Duke of Orléans
General Bonaparte et le Duc d'Orléans

Barye

Models circa 1840's
Listed Barye's Catalogue of 1847

Bronzes illustrated: Dark patination. Exceptional detail. These casts are from the same source as the Candelabra and "Roger et Angélique" sold by Sotheby. They bear similar 'collection' numerals applied to the bronze base in gilt, No. XLIX and XLIII, also the stamps Barye 2 and Barye 4 respectively.

Two exceptionally fine portrait models. Note the 'classical' treatment of the horses, evolved once more from the "Turc".

Example of these works in possession of:-

        Louvre Museum – Paris (Bonaparte only)
        Petit Palais, (Le Duc only)
        (both in Louvre Barye Exhibition 1956/7).
        Walters Art Gallery – Baltimore.
        Corcoran Gallery – Washington.

  Illustration – author's collection.

*Value: rare, about £1,300 for the Duke,
more for the very 'commercial' Napoleon.*

Part-bred Horse
Un Cheval Demi-sang

Barye

Model Circa 1838

Listed in Barye's first catalogue of the 1st September, 1847. Listed in Barye's catalogue of 1855 as new model (larger, about 7½ inches).

Bronze illustrated: Mid-green brown patination. Very fine cast. Height 5½ inches.

This is an early subject by Barye — one of the many examples of his attractive small models of animals in the 'natural' vein.

Examples of this work in the possession of:-

        Louvre Museum (Barye exhibition 1956 — 1957).
        Brooklyn Museum of Art.
        Fogg Art Museum — Cambridge, U.S.A.
        Joslyn Art Museum — Omaha.
        Princeton University Art Museum.
        Wadsworth Atheneum — Hartford, Conneticut.
        Walters Art Gallery — Baltimore.

                Value: Rare £550—£850

Part-bred Horse, Head Lowered
Un Cheval Demi-sang, la Tête Baisée

Barye

Model Circa 1838
Listed in Barye's catalogue of 1847, (also in his catalogue of 1855 as a new model, larger size).

Bronze illustrated: Rich brown patination. Fine detailed cast. Height 5 inches.

Modelled as a pair to previous group.

Examples of this work in the possession of:-

>Brooklyn Museum of Art.
>Corcoran Gallery – Washington.
>Philadelphia Museum of Art.
>Walters Art Gallery – Baltimore.

Value: Rare – £500–£800

Percheron *Cheval Percheron* **Barye**

Model Circa 1850's.

Bronze illustrated: Light bronze patination. Very fine, well finished cast by Fumière et Thiebaut. (An example of sculptor/founders co-operating with Barye). Height 16 inches.

This model (smaller) was listed by Barye as a new work in his catalogue of 1855. Possibly a private commission originally, it has the proud 'romantic' flavour of Barye's 'Turc' – the recognised inspiration for almost all his horses – and none of the 'naturalism' suggested in the previous two models.

Value: Rare   £1,900–£2,500

Orang-Utang Riding a Gnu *Un Orang-Outang Monté sur un Gnou* **Barye**

Model Circa 1840

Listed in Barye's first catalogue of the 1st September, 1847.

Bronze illustrated: Medium brown patination. Fine quality cast with good detail. Height 9 inches.

Unsual subject (perhaps Barye saw this strangley assorted couple at the zoo?).

Examples of this work in the possession of:-

        Louvre Museum (Barye exhibition 1956–1957).
        Corcoran Gallery – Washington.
        Walters Art Gallery – Baltimore.

Illustrated – Saunier.

Value: Rare – £950–£1,500

Lapith and Centaur (A)   Barye
Lapith Combattant le Centaure

Model listed Barye's catalogue of 1855

Bronze illustrated: Rich dark brown over black patination. Signed A-L BARYE. (This form of Barye's signature is usual on casts of this model). Height 14½ inches.

Example of this work in possession of:-

    Victoria and Albert Museum – London. (Purchased at the Paris Exhibition of 1855).
    Walters Art Gallery – Baltimore. (See illustration opposite page).

Value: £1,400–£1,900

Illustration — author's collection.

Lapith and Centaur (B)
Lapith Combattant le Centaure

Barye

Bronze illustrated: This cast is the possession of the Walters Art Gallery. Height 13½ inches. Compare these two casts of an identical subject and note the small differences that occur from one to the other.

Illustration — courtesy of the Walters Art Gallery, Baltimore.

Author's note:
The model for the casts illustrated here is listed in Barye's catalogue as an "Equisse du même sujet" of a much larger, and slightly different version, entitled "Theseus Slaying the Centaur Bièenor" which was exhibited by Barye, in plaster, at the Salon of 1850. In 1877 this plaster was used to cast a bronze by 'cire perdue' which is now in the Louvre. (Barbédienne also cast this larger model — illustrated Ballu).

Theseus Combating the Minotaur　　　　　　　　　　　　　　　　　　　　　　　　　**Barye**
Thesée Combattant le Minotaure

Listed **Barye's catalogue** of 1847.

Bronze illustrated: Dark green-black patination. Good cast by F. Barbédienne (additional gold seal – collection Barbédienne – can be seen in the foreground.) Height 17 inches.

A celebrated work by Barye in the classical manner. Although documented as earlier, it is similar in form to the "Lapith and the Centaur" – note the stylised treatment of the head of Theseus.

Example of this work in possession of:-

      The Louvre Museum, Paris.
      Victoria & Albert Museum, London, (purchased Paris Exhibition 1855).
      Museum of Art, Brooklyn.
      Yale University Museum, New Haven.
      Walters Art Gallery, Baltimore.
      Corcoran Gallery, Washington.

Illustrated – Ballu,
         Saunier.

Value: £800–£1,600

Hercules and the Erymanthean Boar
Hercule Tuant le Sanglier d'Erymanthe

Barye

Model exhibited Salon of 1823

Bronze illustrated: Dark patination. Good detailed cast. Height 5½ inches.

A well-known subject in sculpture, this is an early work by Barye modelled during his time with the goldsmith, Fauconnier.

Example of this work in possession of:-

    The Louvre Museum, Paris.
    Walters Art Gallery, Baltimore.

Value: Rare £180—£300

Charles VIIth Victorious   **Barye**
Le Roi Charles VII Victorieux

Model exhibited Salon of 1832
Listed Barye's 1847 catalogue.

Bronze illustrated: Gold-bronze patination. Exceptional detail (crown of leaves missing). Height 9 inches approx. excluding base).

Early model by Barye. One can see the origin of the "Turc" here.

Illustrated – Ballu

Example of this work in possession of:-

        Museum of Fine Arts, Bordeaux.

                    Value: Rare £850–£1,100

Caucasian Warrior (Equestrian) **Barye**
Guerrier du Caucase

Listed Barye's 1855 catalogue.

Bronze illustrated: Dark black brown patination. Detail could be crisper. Height 7½ inches.

This group was listed (on this scale) as a pair to "Piqueur, (costume Louis XV)". Although catalogued as new models in 1855, they probably originate from the 1840's.

Example of this work in possession of:-

        The Louvre Museum, Paris. (Barye Exhibition 1956/7).
        Victoria & Albert Museum, London. (Piqueur only, smaller).
        Corcoran Gallery of Art, Washington.
        Walters Art Gallery, Baltimore.

              Value: £200−£400

Lion Crushing a Serpent **Barye**
Lion au Serpent

Model exhibited Salon of 1833 (plaster dated 1832).
Listed Barye's catalogue of 1847.

Bronze illustrated: Rich dark patination. Exceptional cast. Stamped Barye 31. Height 10 inches.

One of Barye's outstanding works. This sculpture needs no further praise. The lifesize Salon plaster was bought by the State and cast by Honoré Gonon in 'lost wax'. This was shown in the Salon of 1836 and later placed in the Tuileries Gardens. It can now be seen in the Salle de Barye at the Louvre.

Other examples of this work in Museums or public places include:-

        Baltimore Museum of Art, (Lucas Collection).
        Brooklyn Museum of Art.
        Fogg Art Museum, Cambridge, U.S.A.
        Metropolitan Museum of Art, New York.
        Corcoran Gallery, Washington.
        Walters Art Gallery, Baltimore.
        Ritterhouse Square – Philadelphia (Life size cast by F. Barbédienne)

Illustrated – Ballu
              Saunier.
                 Value: £600–£1,600

Variation of Lion and Serpent **Barye**

Model circa 1840's

Bronze illustrated: Dark green patination. Good surface detail. Height 7 inches approx.

This is a variation of the original "Lion and Serpent". A new model would have been created as there are distinct differences. The positioning of the lion is lower and there are additions to the base, giving it more height (a rock under the body of the snake etc.).

Example of this work in possession of:-

    Victoria & Albert Museum, London.
    Baltimore Museum of Art, (Lucas Collection).
    Joslyn Museum of Art, Omaha.
    Wadsworth Atheneum, Hartford, Connecticut.
    Walters Art Gallery, Baltimore.

Value: £250–£450

Lion Crushing a Serpent with its Hindpaw
Lion la patte Levée sur un Serpent

Barye

Model circa 1830's

Bronze illustrated: Dark patination. Unfortunately the cast illustrated is not the best example of this work. Height 5¼ inches.

Although a 'variation on the theme' of the "Lion and Serpent" this is a different and equally dramatic model.

Example of this work in possession of:-

        Victoria & Albert Museum, London.
        Baltimore Museum of Art, (Lucas Collection).
        Boston Museum of Fine Arts.
        Brooklyn Museum of Art.
        Metropolitan Museum of Art, New York.
        Corcoran Gallery, Washington.
        Walters Art Gallery, Baltimore.

Illustrated – Ballu

Value: £300–£450

Jaguar Devouring a Hare **Barye**
Jaguar Devorant un Lièvre

Model exhibited Salon of 1850 (plaster)

Bronze illustrated: Pale bronze patination. Cast by F. Barbédienne. Signed A — L BARYE *(This signature, with the addition of the initials appears seldom, but it is quite correct and not to be confused with A. Barye — Alfred, Barye's son).* Length 16 inches.

This was one of two models sent to the Salon of 1850; Barye's protest against the 'powers that be' had ended, and after entering nothing for over 10 years he exhibited this sensational sculpture. A work of great dramatic force on a favourite theme — the oppressor and the oppressed — it was another milestone in his career. Barye's practice of accentuating certain parts of the body for emphasis (the head and muscular back legs) was adopted by Rodin (e.g. The Burghers of Calais). Rodin attended Barye's classes when he was Professor at the Natural History Museum.

Example of this work in possession of:-

        The Louvre, Paris (plaster, also large scale bronze).
        Bonnat Museum, Bayonne.
        Walters Art Gallery, Baltimore.
        William Rockhill Nelson Gallery, Kansas City.
        Palace of the Legion of Honour, San Francisco.
        Corcoran Gallery, Washington.

Illustrated — Ballu.
        Saunier.   *Value: £350—£650 size illustrated.*

Note: Excerpt from Louvre catalogue of Barye Exhibition to 1956/57, Exhibit 4. "The plaster model shown at the Salon of 1850 is now in the Louvre (Zoubaloff bequest). A bronze was commissioned by the State on the 8/10/1851 and shown at the Salon of 1852. The bronze shown is the 'modèle' kept by Barye which passed later to the collections of Goupil, Barbédienne and Zoubaloff. Donated to the Museum by Zoubaloff in 1912."

Seated Lion   **Barye**
Lion Assis

Bronze exhibited Salon of 1847
Listed in Barye's first catalogue of 1847

Bronze illustrated: Rich dark patination. Very well detailed small cast. Height 8 inches.

A proud and dignified study. The original model cast by Barye in 'cire' was commissioned for the Tuileries Gardens in 1846. Moved to the Quai-side, at the Louvre in 1867. (Placed on the right, a pair was cast which Barye did not care for).

Example of this work in possession of:-

        The Louvre, Paris (plaster dated 1847).
        Victoria & Albert Museum, London.
        Baltimore Museum of Art, (Lucas Collection).
        Brooklyn Museum of Art.
        Fogg Art Museum, Cambridge, U.S.A.
        Joslyn Art Museum, Omaha.
        Walters Art Gallery, Baltimore.
        Wadsworth Atheneum, Hartford, Connecticut.
        Corcoran Gallery, Washington.
        Academy of Arts, Honolulu.
        Large cast by F. Barbédienne in Mount Vernon Square, Baltimore.

Illustrated – Saunier.
        Antiques International
                Value: £350–£550

Panther of India Barye
Lionne Couchée, le Cou Ramessé

Listed in Barye's catalogue of 1855

Bronze illustrated: Medium brown patination. Clear detail where intended by Barye. Very fine cast. Length 10½ inches.

Called ' with the thick neck '. Barye has accentuated this part particularly on this ' majestic ' model giving lesser detail elsewhere.

Example of this work in possession of:-

    The Louvre Museum, Paris.
    Victoria & Albert Museum, London.
    Baltimore Museum of Art, (Lucas Collection).
    Museum of Fine Arts, Boston.
    Brooklyn Museum of Art.
    Fogg Art Museum, Cambridge, U.S.A.
    Walters Art Gallery, Baltimore.
    Corcoran Gallery – Washington.

        Value: £400–£600

Panther of Tunis *Barye*
Panthère de Tunis

Listed in Barye's catalogue of 1855

Bronze illustrated: Brown rubbed brass patination. Poor cast. Height 3¼ inches.

Example of this work in possession of:-

    Victoria & Albert Museum, London.
    Fogg Art Museum, Cambridge, U.S.A.
    Walters Art Gallery, Baltimore.

                Value: £100–£250

Walking Leopard and Walking Panther   Barye
Un Léopard, une Panthère

Listed as bas-reliefs in Barye's 1847 catalogue.
Circa 1830

Bronze illustrated: *Rich mid-brown patinated bronze plaques, mounted on marble. Exceptional detail. These casts date from the 1830's. The Leopard is No. 11 and the Panther No. 6 (dated 1831). Overall measurements 4½ inches x 13 inches.*

These early works are two of a set of four bas-reliefs, (the "American Stag" and the "Genette Carrying away a Bird").

Example of these works in possession of:-

        The Louvre Museum, Paris. (Exhibited Barye Exhibition 1956/57).
        Baltimore Museum of Art, (Lucas Collection).
        Boston Museum of Art
        Brooklyn Museum of Art.
        Joslyn Art Museum, Omaha.
        Metropolitan Museum of Art, New York.
        Museum of Art, Philadelphia.
        Corcoran Gallery, Washington.
        Walters Art Gallery, Baltimore.

Illustrated – Ballu
               Saunier

Value: £30–£80 each plaque

Tiger Devouring a Gavial **Barye**
Un Tigre Dévorant un Gavial (Crocodile du Ganges).

Model exhibited Salon of 1831 (plaster)
Listed in Barye's first catalogue of 1847.

Example of this work in possession of:-

        The Louvre, Paris (large bronze cast in 1832 by
           Honoré Gonon and his Sons).
        Bonnat Museum, Bayonne.
        Baltimore Museum of Art, (Lucas Collection).
        City Museum of Art, St. Louis.
        Metropolitan Museum of Art — New York.
        Walters Art Gallery, Baltimore.

Illustrated — Ballu

'Lion with Dead Antelope
Lion Tenant un Guib

Listed Barye's first catalogue of 1847

Example of this work in possession of:-

        Metropolitan Museum, New York.
        Walters Art Gallery, Baltimore.

Bronzes illustrated: Unfortunately these two very fine, early casts are poorly illustrated.
        Height 4 inches approx. Lion with Dead Antelope stamped BARYE 7.

        Value: £350—£550 each

Lion
Lion qui Marche
                       Barye

Model circa 1836

Illustration: *Property of the Walters Art Gallery, Baltimore. Cast in silver. Dated 1865. Brass plate on the base (not shown) inscribed "Paris 30th April 1865. Fille de L'Air". Valued at 10,000 f. this was the Grand Prix de Paris, Longchamps Trophy. It was purchased by George Lucas, and passed on to William Walters, on the death of the owner of the winning filly, M. de Lagrange.*

Illustration – Courtesy Walters Art Gallery, Baltimore.

Lion
Lion qui Marche

Barye

Model circa 1840's
Listed Barye's first catalogue of 1847.

Bronze illustrated: Rich black-green patination. This is a good bronze cast, although detail is not as sharp as on the Walters Group.

This fine work originates from the Lion, in relief, for the July Column, Place de la Bastille, erected in 1836.

Example of this work in possession of:-

    Victoria & Albert Museum, London.
    Musée des Arts Decoratifs, Paris.
    (Shown The Louvre 1956/57 Barye Exhibition).
    Baltimore Museum of Art, (Lucas Collection).
    Brooklyn Museum of Art.
    Fogg Art Museum, Cambridge, U.S.A.
    Newark Museum, New Jersey.
    Princeton University Art Museum.
    City Art Museum, St. Louis.
    Wadsworth Atheneum, Hartford, Connecticut.
    Palace of the Legion of Honour, San Francisco.
    Corcoran Gallery, Washington.
    Walters Art Gallery, Baltimore.

Illustrated – Ballu

Value: £300–£450

Tiger
Un Tigre qui Marche

Barye

Model circa 1840's
Listed in Barye's first catalogue of 1847.

Bronze illustrated: Medium dark patination. Very fine cast. Note the detailed stripes on the Tiger. Height 8¼ inches.

Modelled as a pair to the preceding Lion. A powerful, uncluttered work.

Example of this work in possession of:-

        Victoria & Albert Museum, London.
        The Louvre (shown Barye Exhibition of 1956/57).
        Musée des Arts Decoratifs, Paris.
        Baltimore Museum of Art, (Lucas Collection).
        Brooklyn Museum of Art.
        Fogg Art Museum, Cambridge, U.S.A.
        Metropolitan Museum of Art, New York.
        Joslyn Art Museum, Omaha.
        City Art Museum, St. Louis.
        Walters Art Gallery, Baltimore.
        Wadsworth Atheneum, Hartford, Connecticut.
        Honolulu Academy of Arts.
        Portland Art Association, Oregon.
        Palace of the Legion of Honor, San Francisco.
        Corcoran Art Gallery, Washington.

Illustrated — Ballu,
        Saunier.   *Value: £300–£450*

Sleeping Jaguar *Un Jaguar Dormant*   **Barye**

Listed in Barye's first catalogue of 1847.

Bronze illustrated: Rich dark rubbed brass patination. Good cast — there is the rather 'naive' indication of the animal's spots associated with this work. Length 11 inches.

A simple and very effective model.

Example of this work in possession of:-

        Baltimore Museum of Art, (Lucas Collection).
        Brooklyn Museum of Art.
        Metropolitan Museum of Art, New York.
        Corcoran Gallery of Art, Washington.
        Walters Art Gallery, Baltimore.

                Value: *£300–£500*

Lion Devouring a Hind   Barye
Un Lion Devorant une Biche

Model dated 1837.
Listed Barye's first catalogue of 1847.

Bronze illustrated: Silvered bronze. Exceptionally well detailed. One of Barye's stamps can be clearly seen in the foreground. There are two more to the lower left and another on the far right. It is dated 1837. Length 11 inches.

Exceptional model. This is an example of Barye's ability to put a 'twist in the lion's tail', the envy of the painter Delacroix.

Example of this work in possession of:-

>The Louvre Museum, Paris.
>Baltimore Museum of Art, (Lucas Collection).
>Brooklyn Museum of Art.
>Fogg Art Museum, Cambridge, U.S.A.
>Metropolitan Museum of Art, New York.
>Palace of the Legion of Honor, Los Angeles.
>Walters Art Gallery, Baltimore.
>Corcoran Gallery of Art, Washington.

Illustrated — Ballu.

Value: £600—£900

Tiger Devouring an Antelope　　　　　　　　　　　　　　　　　　　　　　　　　　　Barye
Tigre Devorant une Gazelle

Model circa 1830's
Listed Barye's first catalogue of 1847.

Bronze illustrated: Dark green patination. Fine cast by F. Barbédienne. Height 13 inches.

A characteristic example of Barye's imaginative sculpture. This emotional work is among his finest in this sphere – the over-powering of the weaker species.

Example of this work in possession of:-

        Museum of Fine Arts, Boston.
        County Museum of Art, Los Angeles.
        Museum of Art, Philadelphia.
        Metropolitan Museum of Art, New York. (Stone).
        Walters Art Gallery, Baltimore.
        Corcoran Gallery of Art, Washington.

                    Value: £600–£900

Panther Surprising a Civet Cat **Barye**
Panthère Suprenant un Zibet

Listed Barye's catalogue of 1855.

Bronze illustrated: Medium dark patination. Cast most probably dates from the 1840's. Exceptional cast.

A fine model, not 'overloaded', yet projecting the power and strength of the attacker.

Example of this work in possession of:-

    * The Louvre Museum, Paris.
       Baltimore Museum of Art, (Lucas Collection).
       Brooklyn Museum of Art.
       Corcoran Gallery of Art, Washington.
       Walters Art Gallery, Baltimore.

Value: Rare £400–£600

* *Note: The original model (Plaster and wax) sold on Barye's death, together with a bronze with slight differences, are in possession of the Museum. (Thome – Thiery bequest).*

Rearing Bull   **Barye**
Un Taureau Cabré

Listed Barye's first catalogue of 1847.

Bronze illustrated: Medium dark, rubbed brass patination. Excellent surface detail. Very fine cast. Height 8½ inches.

Fine sculpture. The strength and dangerous temperament of this subject is very well suggested.

Example of this work in possession of:-

        The Louvre Museum, Paris. (Barye Exhibition 1956/57).
        Brooklyn Museum of Art.
        Walters Art Gallery, Baltimore.

        Value: £400–£550

Bull, Head Lowered
Un Taureau se Défendent

Barye

Listed Barye's catalogue of 1855.

Bronze illustrated: Rich, dark green patination which does much for this cast which has not the surface detail of preceding group. Height 7 inches.

An excellent character study of this powerful species. Possibly intended as a pair to the former bull.

Example of this work in possession of:-

    The Louvre Museum, Paris. (Exhibited Barye Exhibition 1956/7).
    Victoria & Albert Museum, London.
    Philadelphia Museum of Art.
    Metropolitan Museum of Art, New York.
    Corcoran Gallery of Art, Washington.
    Walters Art Gallery, Baltimore.

Value: £350–£500

Water Buffalo   **Barye**
Buffle

Listed Barye's 1855 catalogue.

Bronze illustrated: Rich brown patination. Good cast by F. Barbédienne. (Additional initials FB in gold). Height 6 inches.

Very unusual subject matter. Original probably dates from the 1830's.

Example of this work in possession of:-

        Baltimore Museum of Art, (Lucas Collection).
        Philadelphia Museum of Art.
        Corcoran Gallery of Art, Washington.

                Value: Rare £300—£450

Stag, Doe and Fawn *Barye*
Cerf, Biche et Faon

Listed in Barye's catalogue of 1847.

Bronze illustrated is the property of the Victoria and Albert Museum. Cast by F. Barbédienne. Acquired by the Museum in 1890. Height 8 inches approximately.

This well composed small group probably dates from the 1830's.

Illustration – courtesy of the Victoria and Albert Museum, London.

Deer and Fawn Barye

Bronze illustrated is the property of the Victoria and Albert Museum. Rich brown patination. Cast by F. Barbédienne. Acquired by the Museum in 1890. Length 5 inches.

Probably the "Faon et Daim" listed in Barye's 1855 catalogue, it is of the period of the previous group.

Illustration — courtesy of the Victoria and Albert Museum, London.

Mule   Barye

Bronze illustrated is the property of the Victoria and Albert Museum. Cast by F. Barbédienne – stamped F.B. gold. Mid-green over brown patination. Height 8 inches approximately. Acquired by the Museum in 1882.

Not listed in Barye's catalogues. This model is probably circa 1860–1870. Rare.

Illustration – courtesy of the Victoria and Albert Museum, London.

Walking Jaguar
Jaguar qui Marche
                                                                                                                                Barye

Listed in Barye's catalogue of 1847.

Bronze illustrated is the property of the Victoria and Albert Museum. Green-brown patination. Cast by F. Barbédienne. Acquired by the Museum in 1890. Length 7 inches.

The date of the model is 1840.

Illustration – courtesy of the Victoria and Albert Museum, London.

Christophe FRATIN 1800 - 1864

Fratin was the son of a taxidermist. He was born in Metz, where he began his studies originally under the sculptor Pioche, who had returned there after having had considerable success in Paris. Later, Fratin was accepted as a pupil by the celebrated painter, Géricault, and made Paris his home. There are no lengthy biographies of his life and work but, a regular contributor to the salon until a year before his death, he made a successful career for himself.

Although Fratin edited most of his sculpture, he had no actual foundry of his own, relying on others to 'pour' the bronze, which could be why a considerable amount of his work was executed in plaster. Some of these plasters were produced by the Susse Frères in the 1830's and it is recorded that he held three large sales of sculpture during his career — the one in March 1857 was for terra-cotta entirely and contained seventy works, all of which were sold. The fragility of these materials explains the fact that there is less work surviving by Fratin in comparison with that of Barye or P.J. Mêne, in spite of his reasonable output of small sculptures. The bronzes that are extant are, almost without exception, of excellent quality — mostly these were cast at his own expense in the workshop of the Quesnels, but later in his life some of his work was in the hands of the founder Daubré.

Exhibiting for the first time at the Salon of 1831, when Barye showed his famous "Tiger and Gavial", Fratin's entry is listed as "Fermer, an English Thoroughbred" — it is interesting that it should be a horse and one of English origin — his sculptures of horses were to be amongst his best subjects, and his subsequent popularity in England a great help to his career. Fratin had considerable success in his own country, winning prizes at exhibitions and receiving private commissions. He was given much encouragement by the various 'departments' of the State, who commissioned work for public places in Paris and particularly for his birthplace of Metz. Public bodies were augmenting the rich patrons abroad as well as in France and Fratin received commissions from Germany, Austria and America. A large group of two eagles and their prey is in Central Park, New York City, and the Peabody Institute, Baltimore, own eleven of his smaller sculptures. Although rebuffed occasionally by the French Salon, his success as a sculptor was already assured. He had influential patrons in England, too, and at the Great Exhibition of 1851 the citation accompanying the medal that he was awarded there described Fratin as the greatest animal sculptor of his day.

With Barye and Mêne, Fratin has an important position in animalier sculpture, and whilst both these artists had imitators, Fratin has none; he was a complete individualist with a very distinct style of his own. In a group of animalier sculpture any work by Fratin is immediately recognised but opinion of his individual approach has always been divided. His upholders are invariably sensitive to his art and his disclaimers, students of nature. Occasionally he does slip up, presenting slab-sided animals which would be more at home under the glass dome of the taxidermist. When appraising Fratin's work one does not look too closely for perfection of anatomical detail, handled so simply and unobtrusively by Barye and so competently by Mêne, but for his own very unique treatment of the whole work. The original conception of Fratin's groups is romantic, and although his treatment in no way attempts the complete freedom of line that was Barye's *tour de force,* he sculpted in his own way groups that were vivid and imaginative and which have a rich, almost ragged finish more usual to the palette knife than to the chisel. The answer to Fratin's very distinctive approach to sculpture must lie with his master Géricault. The romantic ideas and the 'impasto' technique of this great painter were translated by Fratin into his own medium, sculpture.

Dancing Bear Fratin

Bronze illustrated: Dark patination. Although 'rugged' and not over detailed, the surface texture is well defined on this cast. Height 7½ inches.

One of Fratin's very unusual compositions.

Value: Rare £150—£250

Dancing Monkey   **Fratin**

Bronze illustrated: *Rich dark brown black patination. Fine cast by Quesnel. Signed Fratin. Height 6¾ inches.*

Interesting and original composition.

Value: Rare £200–£300

Pharoah Hound   **Fratin**

Bronze illustrated: Rich dark brown patination. Very good detail on the cast. Stamped Fratin. Height 6 inches.

Very unusual subject. Excellent model.

Value: Rare £150–£250

Racing Greyhounds — Fratin

Bronze illustrated: Brass patination. An ill defined cast. Height 4 inches. Interesting model, shows much movement.

Value: Rare £75–£150

Hound   Fratin

Bronze illustrated: Overall mid-brown patination which gives rather a dull effect to rather a limp model. Stamped Fratin, also Daubry éditeur. Height 6 inches.
Value: £75—£120

Note: Stamp can be seen in the foreground.

Bronze illustrated: Very dark patination with brass evident. A good crisp cast. Model typical of artist's most interesting work. Height 10½ inches.

Value: Rare £200–£300

Note: Stamped Fratin (the N is often reversed on this small stamp, which can be seen in the foreground of the base).

Setter   **Fratin**

Bronze illustrated: Dark patination. This incredibly 'sharp' cast is among the finest founding examples of this artist's work — inscribed under the base "Premier épreuve". Length 15 inches.

Model rather stiff.

Value: Rare £250—£450 upwards

Lion with Kill — Fratin

Model Exhibited at the Salon of 1835.

Bronze illustrated: Light brown patination. Good cast – shows clearly Fratin's 'rugged' surface detail on the lion. Height 6½ inches.

Not a very inspiring model for a subject of this sort.

Value: Rare £100–£250

Lioness   **Fratin**

Bronze illustrated: Brass brown patination. An excellent quality small bronze. Height 3 inches.

A good small model.

Value: Rare £100—£150

Mare and Foal Fratin

Bronze illustrated: Mid-brown patination. A good cast with evidence of very detailed chiselling — unusual on casts of Fratin's work. Length 17 inches.

Possibly the same model as exhibited by Fratin, in plaster, at the Salon of 1837.

Illustration — Courtesy Sotheby.

Value: Rare £700–£900

Mare and Foal Fratin

Bronze illustrated: Very rich dark patination. A good quality cast. Although less surface detail than the other group illustrated, it is very typical of Fratin's rugged style. Height 9 inches.

Fratin exhibited a plaster model of a Mare and Foal at the Salon of 1837. No descriptive detail is given, but it is probably this, or the other group illustrated.

Value: Rare £650—£850

Bronze illustrated: Rich brown black patination. Good cast. Height 9¾ inches.

Perhaps a portrait group, it is less picturesque than some of Fratin's other sculpture of these subjects.

Value: Rare £600–£900

Horses Attacked by Wolves   Fratin

Bronze illustrated: Brown patination. Excellent cast. Height 17½ inches.
Interesting and well composed 'romantic' sculpture.
Value: Rare £450–£650

Arab Stallion   **Fratin**

Bronze illustrated: Rich dark brown patination. Good quality cast. Treatment of the mane and tail typical of Fratin's very individual style. Height 9 inches approx.

This cast was sold for £750 at Sotheby's first collective sale of animalier sculpture in 1968.

Value: Rare £600–£900

Boar   Fratin

Bronze illustrated: *Brass bronze patination. Sharper surface detail would have been in keeping with this subject — more suggestion of 'bristles' on the coat. Height 3 inches.*

The quick and dangerous character of the animal is well suggested in the modelling.

Value: Rare £50–£100

Bull   **Fratin**

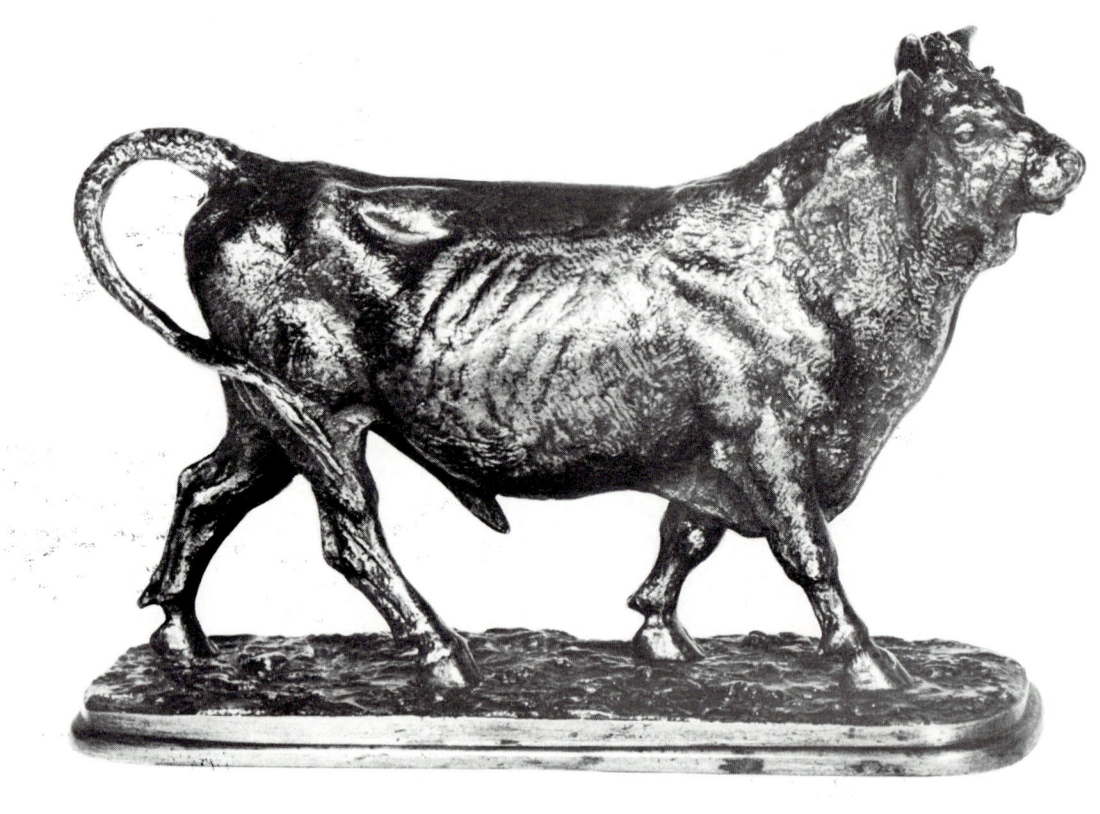

Bronze illustrated: Gold-bronze patination. Over-rubbed surface detrimental to the cast. Height 6 inches.

Value: Rare £50–£150

Bears Playing Fratin

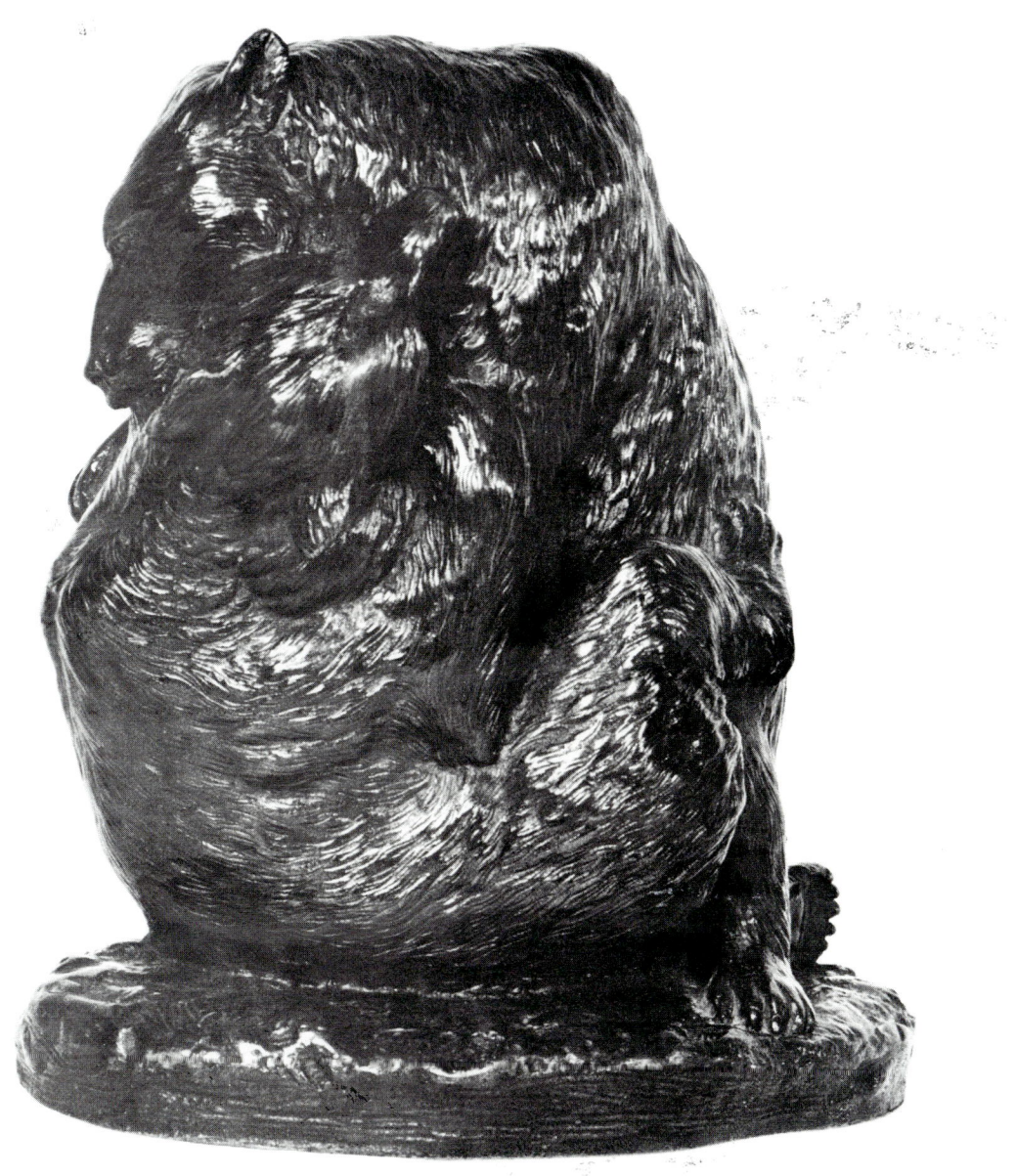

Model circa 1832

Bronze illustrated: Rich dark brown patination. Dated 1832. Signed Quesnel – fondeur. It has the 'waxy' look associated with 'cire perdue' casting. First quality cast. Height 6¾ inches.

The positioning of the two bears is excellent – a very fine composition altogether.

Illustration – author's Collection.

Value: Rare £300–£500

Two Eagles with Prey **Fratin**

Model circa 1835

Bronze illustrated: Medium grey black patination. Very fine chiselling. Founded by Daubré. Height 20 inches.

A well-modelled and interesting composition.

Value: Rare £250–£450

Stags   Fratin

Models Circa 1835.

Bronzes illustrated: Rich black patination. Typical of Fratin's 'texture'. Height 6 inches approx.

Value: Rare £100–£200 each

Model illustrated: Brown patination. Not one of the best casts. Height 16 inches.
Majestic model — good representation of the subject.
Value: £200—£500

Pierre Jules MENE 1810-1879

A Parisian, Mêne seemed to have good fortune throughout his whole career. He was born on the 15th March, 1810, at 84 Rue du Faubourg St. Antoine, and was the son of Dominique Mêne, a metal turner. He was taught the rudiments of sculpture and founding by his father, and when he married at twenty-two, he was earning his living making models for reproduction by the porcelain manufacturers, these models being much in vogue at the time. He received further tuition from the sculptor, René Compaire, and finding a very definite talent for animal sculpture, he went about it thoroughly, spending his time at the zoo, where he could study his subjects and make diligent sketches and studies. His progress was rapid, and after he exhibited at the Salon of 1838 for the first time, his efforts were well rewarded. He received the first class medal twice, and many other awards, including the Cross of the Legion of Honour in 1861. His career was assured, and he was a regular exhibitor at the Salon up to the year of his death.

Mêne lived for part of his life in the Faubourg du Temple and operated his own foundry, moving, around 1857, to 9 Rue de l'Entrepôt. He issued a combined catalogue with his son-in-law, Cain, from this address, where he lived until his death. Talented, industrious and popular with the public, he did not need to solicit support in official circles — only one of Mêne's works "The Mounted Huntsman and his Hounds", was acquired by the State in his lifetime — he was content to live and support his family from his work, and was certainly one of the most prolific and successful sculptors of the animalier school. As a result of the reputation he made for himself, examples of his sculpture are now in the Louvre and other French museums, and the Ashmolean, Oxford, has a particularly interesting example of Mêne's work. Exhibiting in England at the Great Exhibition of 1851 and 1862 — there could have been no doubt of his success there, he was the Landseer of sculpture — it is probable that as much of Mêne's work came to nineteenth century England as remained in France.

Mêne's sculpture is of particular importance when discussing the work of the nineteenth century animalier school today as, working in a style in such direct contrast to Barye, and never to be as renowned in his own lifetime as his contemporary, he has nevertheless emerged after over a century as one of the foremost sculptors of this school. Barye was the initiator, and developed his own style, which had a profound effect on the future of sculpture but it was Mêne who surpassed all in his portrayal of animals in the realist form. He is today the one most associated with and typical of the animalier school as a whole. Although one or two of Mêne's earlier works show the romantic influence of Barye (the "Tiger and Alligator" for example), he discarded this and went his own way, creating sculpture of animals directly from nature. Unposed and very much alive, his subjects are captured in a fleeting movement, picturesque in every detail. Mêne's animals are often individual portraits with 'humanised' personalities — in many cases one knows their actual names.

Two painters particularly influenced Mêne: from Landseer he found not only subject matter, but an expressive element, sentimentality, which is in no way detrimental; from the Frenchman, Carle Vernet, whose work he admired and personally collected, he captured the verve and spirit of the painter's compositions in sculptural form.

Much of Mêne's life, like Fratin's, must be pure conjecture, as there is a dearth of biographical detail available, but it is known that his house in the Rue de l'Entrepôt was a meeting place for his many friends among the sculptors, painters and musicians of Paris who were attracted by his personality and his innate taste for all the arts. Benevolent and good-humoured, he was doubtless a contented man, and this, rather obscurely perhaps, has made Mêne one of the most collected animalier sculptors today. Surrounded as we are by disturbance and even violence, possessors of his work can find relaxation and enjoyment in contemplating the innate grace and beauty of his sculpture.

Author's Note

Mêne's catalogue was issued from No. 9 Rue de l'Entrepôt, where he lived from 1857 onwards. My copy is incomplete, but where possible I have put his own catalogue number against the model concerned.

Jaguar and Alligator
Jaguar et Caiman

Mêne

Model exhibited Salon of 1843 (bronze)

Bronze illustrated: Very rich dark patination. An excellent cast. Height 7 inches.

An early model. The 'romantic' style of this work was rarely seen in Mêne's sculpture.

Value: Rare £400–£600

Walking Stag   **Mêne**

Model circa 1840's

Bronze illustrated: Dark patination. Not a very crisp cast (compare the two following illustrations). Height 6½ inches.

Example of Mêne's many small sculptures of individual types of deer.

Value: £75–£120

Stag Browsing
Cerf Broutant (Cerf à la Branche)

Mêne

Model Circa 1843
No. 118 Mêne's Catalogue

Bronze illustrated: Dark patination. Excellent well detailed cast. Height 14½ inches.

A well modelled interesting work.

Example dated 1843 in possession of the author.
Example of this work in possession of Musée de la Rochelle.

Value: £200–£380

Stag Attacked by Three Hounds
Chasse au Cerf

Mêne

No. 21 Mêne's Catalogue
Model exhibited Salon of 1844 (bronze)

Bronze illustrated: Dark, rubbed brass patination. Superbly finished. This is an exceptional cast altogether. Dated 1844. Height 12 inches.

Excellent work. Mêne was able to model intricate groups composed of many animals without overloading the sculpture.

Value: Rare £300—£450

Battling Stags   **Mêne**

Model Exhibited Salon of 1833 (wax)
(Donated to the Louvre Museum by Mme. Cain).
No. 116 Mêne's Catalogue

Bronze illustrated: Mid-brown patination typical of the Susse Frères. A good cast. Height 6½ inches.

Well portrayed attitude — good realist sculpture.

Value: £400—£600

Roebuck   **Mêne**
Chevreuil

Model Exhibited Salon 1861 (bronze)

Bronze illustrated: Brown, rubbed brass patination. Not very detailed. No chiselling or afterwork apparent. Height 7½ inches.

Well modelled example – one of several single studies of the species by Mêne.

Value: Rare £250–£425

Algerian Gazelle　　　　　　　　　　　　　　　　　　　　　　　　　　　　**Mêne**

Model circa early 1840's

Bronze illustrated: Rich dark patination. Fine detail and chiselling. Inscribed "Gazelle femelle de l'Algérie". Height 4¼ inches.

These small models, with species named, are early examples of Mêne's work – much collected.

Value: Rare £100–£150

Leaping Chamois　　　　　　　　　　　　　　　　　　　　　　　　Méne
Chamois Sautant

Bronze illustrated:　Medium dark patination. Exceptional surface detail. Dated 1850. Height 8¼ inches.

Great movement in this model. A pity the necessary support was not better contrived.

　　　　　　　　　　Value: Rare £180–£250

Pair of Roe Deer **Mêne**
Chevreuils

Model Exhibited Salon of 1859 (wax)

Bronze illustrated: Rich bronze patination. More surface detail can be seen on this than on the single Roebuck illustrated. A good cast. Dated 1859. Height 10¾ inches.

Well composed group. A good study of this particular species.

Illustration – Courtesy of Sotheby & Co.

Value: Rare £200–£360

Rooster **Mêne**

Bronze illustrated: Property of the Walters Art Gallery. Unusual early model by Mêne. Height 5½ inches.

Illustration – Courtesy of the Walters Art Gallery, Baltimore.

Fox and Dead Cock Mêne
Reynard et Coq

Model circa 1841

Bronze illustrated: Brown rubbed brass patination. Good cast. Height 3¾ inches.

The mean 'foxy' character is well interpreted. The model is most likely the Salon exhibit of 1841, entitled "Arctic Fox and Cock".

Value: Rare £140–£240

Pair of Foxes Mêne
Groupe de Reynards

Model circa 1840–1845

Bronze illustrated: Bronze rubbed brass patination. Very fine crisp cast. Height 4 inches.

This and other sculpture of foxes date from this period. All have the flat edge to the base associated with Mêne's early, small models.

Value: £180–£280

Fox in a Snare **Mêne**

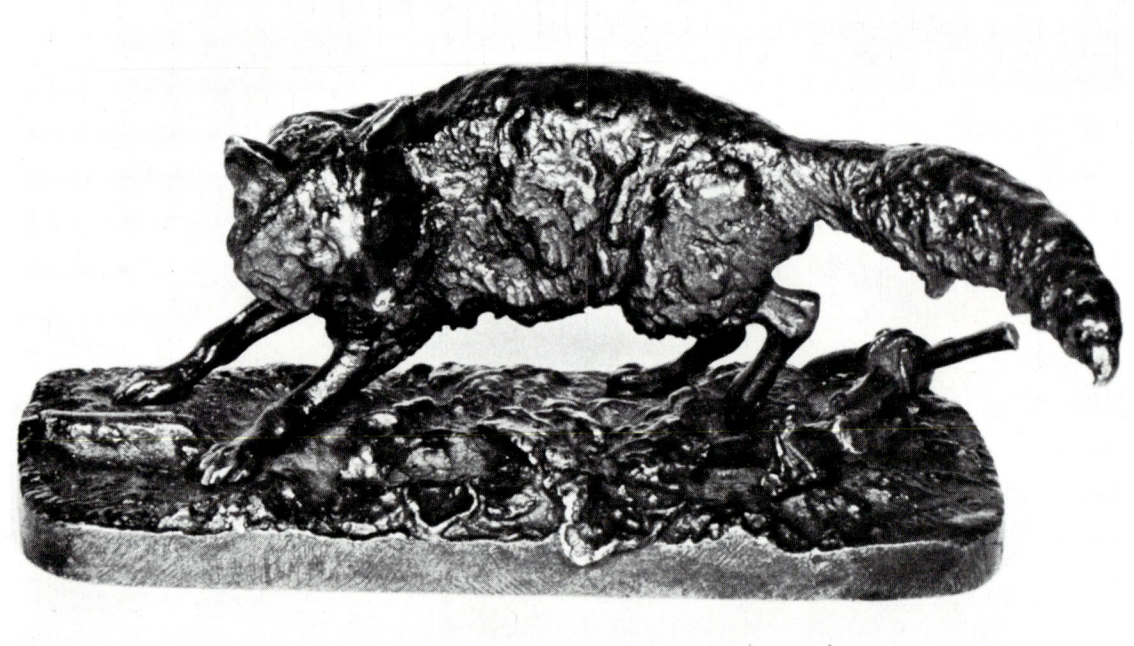

Model Circa 1840–1845

Bronze illustrated: Dark patination. Not a well defined cast. Height 3 inches.
Value: £100–£180

Duck and Ducklings *Mêne*
Famille de Canards

Model Circa 1850's

Bronze illustrated: Bronze, rubbed brass patination. Susse Frères cast. Not exceptional, but reasonably good detail. Height 3¼ inches.

Excellent example of realistic attitude.

Value: Rare £100–£250

Two Ducks *Mêne*
Groupe de Canards

Model Circa 1840–1845

Bronze illustrated: Black brown patination. Good detailed cast. Height 2½ inches.

Very characteristic model.

Illustration – Author's collection

Value: Rare, about £100

Family of Goats
Famille de Chèvres

Mêne

Model Circa 1845–1850

Bronze illustrated: Rich dark patination. Excellent detail – Very good cast. Height 5 inches.

One of Mêne's families of animals which all show exceptionally well the character of the subject portrayed.

Value: Rare £200–£320

Nanny Goat and Kid
Chèvre et Chevreau

Mêne

Model Circa 1850

Bronze illustrated: Dark patination. A good bronze cast by the English foundry Coalbrookdale. Stamped Coalbrookdale bronze. Height 9¾ inches.

A well modelled characteristic study

Value: Rare £100–£200

Rabbits
Groupe de Lapins

Mêne

Model Circa 1850

Bronze illustrated: Brown, rubbed brass patination. Mediocre cast. Height 3¾ inches.
Good composition – very popular subject.

Value: Rare £150–£250

Bull
Taureau Normand

Mêne

Model exhibited Salon of 1845 (bronze)
No. 137 Mêne's Catalogue

Bronze illustrated: Medium dark rubbed brass patination. An exceptionally fine cast, it was presented as a trophy for Agriculture by the Duc d'Aumale. Height 9¼ inches.

Fine model showing the domestic, as opposed to the fighting variety.

Example of this work in possession of the Louvre Museum, Paris, (wax), dated 1844.

Value: Rare £400–£600

Boar Attacked by Hounds
Chasse au Sanglier

Mêne

No. 19 Mêne's Catalogue

Bronze illustrated: Silvered bronze dated 1846. Exceptional cast with fine chiselled detail. Height 10 inches.

One of Mêne's early models. A fine composition (a difficult and controversial subject matter today).

Bronze cast also illustrated — not such fine quality.

Value: Rare £450—£600

Two Whippets at Play
Groupe de deux Levrettes Jouant à la Boule

Mêne

Model Exhibited Salon of 1848 (plaster)
No. 93 Mêne's Catalogue

Bronze illustrated: Rich brown patination. Good detailed cast. Height 6½ inches.

One of the most typical examples of the best of nineteenth century realist 'animalier' sculpture, the composition of this group is superb. It is understandable that it is not a rarity today as a large edition of this popular model was no doubt cast. Both models for this group appear as single sculptures. (The larger one was "Jiji", and the smaller "Gisella").

Value: £150–£350

Greyhound (Italian) and King Charles Spaniel
Group Lévrier et King Charles

Mêne

Model Circa 1848

Bronze illustrated: Dark rubbed brass patination. Not the best of casts. Height 6½ inches.

Modelled as a pair to the previous group. An equally successful and typical work. Both dogs are found as single sculptures.

Illustrated — Antiques International

Value: £200–£380

Greyhound and King Charles Spaniel, and Two Whippets with a Ball — Mêne

Bronze illustrated: Gilded bronze variations of previous groups. Very fine quality. Unusual patination. Height 6½ inches.

Value: Same as bronze

King Charles Spaniel *Mêne*
Chien King Charles

Model circa 1840's

Bronze illustrated: Rich brown patination. Good quality cast. Height 3¼ inches.
The little spaniel was originally modelled for the larger group.

Value: £80–£140

Greyhound Mêne
Lévrier Espagnol

Model exhibited Salon of 1845 (bronze) entitled *"Un Lévrier Espagnol, Grand Espèce"*

Bronze illustrated: Dark, rubbed brass patination. Exceptionally fine quality cast, dated 1844 — straight and unmoulded edge to the base denotes a cast of this period. Height 8¾ inches.

An unusual subject for the artist, it portrays admirably the 'rangy' coursing or racing greyhound as we know it. Compare this to his models which are simply entitled Greyhounds (the whippet, or Italian greyhound of today)

Value: Rare £200–£400
Smaller (about 3½ inches), also rare £60–£120

Setter, Pointer and Partridge
Chasse à la Perdrix

Mêne

Model exhibited Salon of 1848 (wax) 1850 (bronze)
No. 25 in Mêne's Catalogue

Bronze illustrated: Rich dark patination. Excellent quality cast, with exceptional detail. Dated 1847. Height 9½ inches.

Although Mêne did not exhibit the wax of this group until the following year, or show a bronze cast until three years later, 1847 would be the date of the model itself. Both dogs appear as separate works later. The individuality of Mêne's dogs is always apparent, the model for the setter was "Sylphe" and for the pointer, "Tac".

Value: £300–£480

Pointer and Setter Mêne
Groupe Chiens au Taillis

Model circa 1850's
No. 61 in Mêne's Catalogue

Bronze illustrated: Dark black/brown patination. A very fine cast by F. Barbédienne (see top left hand of base). Height 5½ inches.

A lively work similar to the previous illustration. Both models, the pointer "Chien Braque" (Tom) and the setter (very rare), appear as single sculptures.

Value: Rare £300–£450

Irish Setter *Chien Epagneul Anglais (Médor)*   **Mêne**

Circa 1840
No. 73 in Mêne's Catalogue

Bronze illustrated: Rich black patination. Good quality cast. Straight side to the base indicates a cast prior to 1845. Height 6 inches.
Value £120–£250

Pointer – Head Raised Mêne
Chien Braque (Marly)

Circa 1860's
No.77 Mêne's Catalogue

Bronze illustrated: Mid bronze patination, rubbed brass. Nice quality cast. Height 7½ inches.

Another good model of this breed. There is nothing 'wooden' about Mêne's subjects of this sort so often found in the work of his many imitators.

Illustration — Author's collection.

Value: Rare £150–£280

Cropped-eared Terrier with Ball   **Mêne**

Model circa 1860's

Bronze illustrated: Medium dark patination. Good quality cast. Height 4½ inches.

Another obvious portrait model. Perhaps "Lutine" or "Frisette" listed in Appendix D.

A 'wax' is also shown, a dark red (artificially coloured). This would have been used to form a piece-mould and not to be 'lost' in casting. It is now very hard and brittle. (Author's collection).

Value: Rare £80–£180

Bloodhound
Chien Limier

Mêne

Model circa 1860's
No. 67 Mêne's catalogue

Bronze illustrated: Rich brown patination. A very good cast by F. Barbédienne. Height 7 inches.

A typical 'hound' of the period as seen in other larger groups by Mêne.

Value: £90–£180

Pointer — Mêne
Chien Braque Seul

Model circa 1843
No. 78 Mêne's catalogue

Bronze illustrated: Medium dark patination. Fine cast of the period prior to 1848. Height 7 inches.

There are variations of this model, the dog being shown carrying a duck or a hare.

Value: £120–£220

Retriever   **Mêne**
Chien Epagneul Francais (Fabio)

Model circa 1840's
No. 83 Mêne's catalogue

Bronze illustrated: Medium dark patination, rubbed brass. An exceptionally crisp and detailed cast. Prior to 1845. Height 6½ inches.

Fine model, about 1843. Probably cast as a pair to the preceding model.

Value: £150–£250

Pointer Guarding Dead Game   Mêne
Chien Braque, Anglais Pur-sang, Gardant du Gibier

Wax model exhibited Salon of 1850, entitled as above.

No. 58 Mêne's catalogue

Bronze illustrated: Rich dark patination. Very detailed, fine quality ca: Height 12 inches.

A well modelled work.

Note: 'Braques' are pointers, and 'Epagneuls' setters or retrievers, as we know them today.
Value: £300–£450

Two Pointers   Mêne
Groupe Chiens au Repos (Race Saintongeoise)

Circa late 1850's.
No. 57 Mêne's catalogue

Bronze illustrated: Rich brown patination. Good quality cast. Height 10 inches.

Exceptionally fine model, perhaps Mêne Salon exhibit of 1857. A variation of this group, "Chien de Meute avec ses Petits" (Salon 1859), shows the hound on the right of the group with a litter of puppies.

Illustration — author's collection.

Value: Rare £200–£400

Setter Backing
Chien Epagneul (Diane)

Mêne

Bronze illustrated: *Brass bronze patination – not a well detailed cast. (Many of the small examples of dogs seem to lack in this respect). Height 3½ inches.*

One of Mêne's very lively small models.

Value: Rare £60–£120

Cavalier King Charles Spaniel — Mêne

Circa 1840's

Bronze illustrated: Medium, rubbed brass patination. Not particularly good cast. Stamped P.J. Mêne (this way of signing is seen on fairly early models). Height 3 inches.

Another very individual portrait model. Possibly Mêne's Salon exhibit of 1843, "Cross-bred Spaniel".

Value: Rare £80–£120

Three Dogs Burrowing Mêne
Chasse au Lapin

Dated 1853, the wax model was exhibited at the Exposition Universelle of 1855. Exhibited Salon of 1872 (bronze).

No. 24 in Mêne's catalogue

Bronze illustrated: Rich dark patination. Fair quality cast. Height 8 inches.

One of Mêne's most popular and successful "Hunts". This is intricate, but not cluttered, and expresses the eager searching and intelligence of the subjects admirably.

Examples of this work are in possession of:-

        The Louvre Museum — Paris. (wax) Donated by Mme. Cain.
        Museum of Marseilles.
        Carnavalet Museum, Paris.

Illustration — author's collection.

Value: £150—£280

Two Setters and Mallard　　　　　　　　　　　　　　　　　　　　Mêne
Chiens Epagneul – Griffon Saissisant au Canard

Model exhibited Salon of 1850 (bronze) "Chasse au Canard".
No. 22 in Mêne's catalogue.

Bronze illustrated: Rich medium dark patination. A very fine cast with excellent crisp detail. Length 17½ inches.

Perhaps not a favourite subject today. This very spirited model was popular enough in the nineteenth century to be reproduced in England by Coalbrookdale.

Illustration – courtesy of Sotheby & Co.

Value: Rare £200–£350

Pointer
Chien Braque (Tom)

Mêne

Model circa 1850's
No. 96 in Mêne's catalogue.

Bronze illustrated: Pale bronze patination. Fairly well-detailed cast. Height 5 inches.
Value: £90–£180

Mare and Foal
Jument Normande et son Poulain

Mêne

Model exhibited Salon of 1868 (wax), 1869 (bronze).
Exposition Universelle 1878.
No. 30 in Mêne's catalogue.

Bronze illustrated: Rich brown patination. Fine cast. Dated 1868. Founded by F. Barbédienne. The only two examples seen of this large group by Mêne were both cast by this founder. Height 18 inches.

Superb realist model. Excellent character study of both Mare and Foal.

Value: Rare £1,800–£2,500

Mare
Jument Normande Seule

Mêne

Model 1868.
No. 33 in Mêne's catalogue.

Bronze illustrated: Rich dark brown patination. Excellent cast by F. Barbédienne. Height 18 inches.

Related to previous group.

Value: Rare £500–£850

148

Mare and Dog　　　　　　　　　　　　　　　　　　　　　　　　　　　　　　**Mêne**
Jument à l'Ecurie Jouant avec un Chien

Model exhibited Salon of 1859 (bronze).
No. 34 in Mêne's catalogue.

Bronze illustrated: Gold bronze patination. Fairly good detail on cast. Height 9 inches.

This model, christened by someone, sometime, "The Good Companions", shows the sculptor in a 'sentimental' mood portraying a subject associated with paintings of the period.

Value: £400–£650

"Winner of the Race" — Mêne

Model exhibited Salon of 1866 (wax).

Model illustrated: *This wax, the property of the Ashmolean Museum, Oxford, is dark reddish brown in colour, with the addition of a binding agent, probably plaster of Paris. The model is supported internally with wire and would have been used to form a piece-mould, not 'lost'. Inscribed "Lad faisant boire du champagne à un cheval de course tenu par son jockey". Definitely an unfinished model, it is dated 23rd October 1861. As far as it is known this group was never cast in bronze.*

This may be Mêne's actual Salon exhibit. It is certainly a unique piece purchased by the Museum in 1962.

Illustration – courtesy Ashmolean Museum, Oxford.

Dismounted Jockey and Racehorse
Vainqueur de Derby

Mêne

Model Exhibited Salon of 1863 (wax) 1864 (bronze).
No. 13 in Mêne's catalogue.

*Bronze illustrated: Rich brown patination. Fine cast dated 1866. Inscribed "Vainqueur!!!"
Height 14 inches.*

One of Mêne's exceptional models.

Example of this work in possession of:-

> The Heritage Foundation – Canada.
>
> *Value: Rare £1,000–£2,000*

Horse and Jockey　　　　　　　　　　　　　　　　　　　　　　　　　　　　Mêne
Jockey à Cheval

Model circa 1863.
No. 16 in Mêne's catalogue.

Bronze illustrated:　Rich dark patination – very fine cast. Dated 1863. Additional brass plaque inscribed "Vainqueur du Derby". Jockey's whip missing. Height 16½ inches.

One of Mêne's most beautiful compositions.

Illustrated – Antiques International.

　　　　　　　　　　Value: £1,000–£2,000
　　　　　　　　　　Known to have been recast.

Horse and Jockey — Mêne

Bronze illustrated: A rich brown patination. Although not so finely detailed as the preceding one, this is a good cast by F. Barbédienne. Whip missing.

Photograph by courtesy of Henry Spencer & Sons, Retford.

Value: £500–£850

Arab Stallion (Ibrahim)
Ibrahim, Cheval Arabe Ramené d'Egypte

Mêne

Model exhibited Salon of 1843 (bronze).
No. 47 in Mêne's catalogue.

Bronze illustrated: Brass patination. Good crisp cast. "Ibrahim" inscribed on front of base. Height 12 inches.

Good portrait group.

Value: £500–£800

Barbary Stallion (Djinn)
Cheval à la Barrière (Djinn)

Mêne

Model 1848. Exhibited Salon of 1849 (bronze).
No. 43 in Mêne's catalogue.

Bronze illustrated: Rich dark patination. Good cast – not exceptional. Height 11½ inches.

The wild 'calling' horse is well done and the sculptor has managed the barrier adequately. Perhaps a little over-intricate, it is nevertheless far better to see than a solid mass obstructing the horse. ("Djinn" is sometimes tampered with – removed entirely from the original base, or the barrier taken off).

Value: £400–£750

Britanny Stallion
Cheval Breton

Mêne

Model Exhibited Salon of 1863 (bronze).
No. 45 in Mêne's catalogue.

Bronze illustrated: Rich dark patination. An exceptional cast. Height 10 inches approx.
A fine and rare model.

Value: £800–£1,200

Arab Horse Saddled
Cheval de Spahi au Piquet

Mêne

Circa 1850's.
No. 41 in Mêne's catalogue.

Bronze illustrated: Medium dark patination rubbed brass with handling. A very fine cast indeed. Height 11½ inches.

This model appears in other groups by Mêne.

Value: Rare, about £500

Arab Horse Tethered
Cheval au Palmier

Mêne

No. 42 in Mêne's catalogue.

Bronze illustrated: Rich mid-brown patination. Quite a good cast by the Susse Frères. The surface detail could be sharper. Height 11¼ inches.

Good model showing the character of this breed.

Value: £350–£650

Moroccan Horseman
Chasseur Africain

Mêne

Model exhibited Salon of 1878 (wax), 1879 (bronze).
No. 8 in Mêne's catalogue.

Bronze illustrated: Rich brown patination. Very fine and detailed cast.

Impressive group. Mêne has adapted his "Arab Horse Tethered" for this model.

Value: £300–£600

Mounted Huntsman with Hounds at Bay
Groupe Chiens en Defaut (Valet Louis XV).

Mêne

Circa 1860's.
No. 6 in Mêne's catalogue.

Bronze illustrated: Medium dark patination. Quite a good cast. Surface detail could be crisper. Height about 14 inches.

Well composed group — unusual and seldom seen.

Illustration — courtesy of Sotheby & Co.

Value: Rare £300–£500

Louis XVth Mounted Huntsman
Veneur Louis XV à Cheval

Mêne

No. 5 in Mêne's catalogue.

Bronze illustrated: Rich bronze patination. The mane and tail of the horse, also part of the huntsman's clothing and the saddlery etc., are gilded. Probably executed as a special commission or for an exhibition. An exceptionally fine cast. Height 18 inches approx.

A supremely elegant composition.

Value: Rare £1,000–£2,000

Arab Mare and Stallion
Groupe Chevaux Arabes (L'Accolade)

Mêne

Model circa 1850's
No. 27 in Mêne's catalogue.

Bronze illustrated: Rich dark brown-black patination. Superb chiselling and general detail. Exceptional cast. Height 13 inches.

This group was one of Mêne's finest and most successful sculptures. It was exhibited in wax at the Salon of 1852, in bronze in 1853, and at the Exposition Universelle of 1855. Originally entitled "Tachiani et Nedjibé, Arab horses", both these subjects have been used by the artist in other groups. The stallion appears on his own, also Nedjibé. Showing the individual characteristics superbly, the spirited stallion and the very typical 'tetchy' mare, the composition of this group is exceptional.

Example of this work in possession of:-

    The Louvre Museum, Paris.

        Value: £800–£1,800

Arab Mare and Stallion Mêne

Bronze illustrated: A larger edition of the preceding. Height 17 inches — any examples known dated 1865. The detail is not as delicate as on the smaller sculpture.

Value: Rare in this size £800–£1,800

Note: There is a considerably smaller model of this group (height about 8 inches) which is sought after because of its accommodating size. The value is much the same as this very large edition.

Arab Mare and Stallion Mêne

Bronze illustrated: This cast by Susse is quite good, but does not compare in detail with the one illustrated opposite. The putty-coloured brown shade of patination very favoured by this foundry can be seen as much lighter when compared to the others illustrated. Height 13 inches.

Value: £600–£900

Arab Stallion
Cheval Libre

Mêne

Model circa 1850's
No. 39 in Mêne's catalogue.

Bronze illustrated: A rich brown/black patination. Excellent surface detail. A fine quality cast. Height 11½ inches.

A spirited model. The stallion as shown in the group of the "Arab Mare and Stallion".

Value: Rare £600–£850

Racehorse Saddled and Bridled   Mêne
Cheval de Course

Circa 1860's
No. 35 in Mêne's catalogue

Bronze illustrated: Light bronze patination – Good quality cast. Height 9½ inches approx.

A variation of Mêne's "Horse and Jockey". Good realist work, typical of this artist.

Value: Rare £700–£950

Arab Mare, Saddle and Gun at her Feet   Mêne
Jument Arabe (Nedjibé) avec Harnachement

Circa 1850
No. 38 in Mêne's catalogue.

Bronze illustrated: Dark patination. Good cast. Height 11 inches.

The mare is modelled in the group of the "Arab Mare and Stallion".

Value: Rare £500–£750

Mare and Foal   **Mêne**
Jument Arabe et Son Poulain.

Model exhibited Salon of 1850.
No. 31 in Mêne's catalogue.

Bronze illustrated: Bronze patination. A reasonably good Susse Frères cast, perhaps not as finely detailed as some. Height 10½ inches.

The wax model exhibited by Mêne at the Salon was entitled "Kemkon – Handani Arab Mare and Foal". This is fine realist sculpture showing the characteristic mare-foal relationship – the irritable mare with an over-playful offspring.

Value: £900–£1,500

A small edition of this sculpture (height 6 inches) was cast.

Value: Rare £600–£850

After the Hunt in Scotland
"La Prise du Renard – Chasse en Ecosse"

Mêne

Wax model exhibited Salon of 1861. Shown, in silver, at the Great Exhibition of 1862 and the Exposition Universelle of 1867.
No. 2 in Mêne's catalogue.

Bronze illustrated: Dark bronze patination. Excellently cast by F. Barbédienne. Dated 1861. Height 20½ inches.

This work shows the influence of the painting of Sir Edwin Landseer, the great favourite of Victorian England (France, too, rewarded him, with the Gold Medal at the Exposition Universelle of 1855). This large and intricate work must have been a considerable challenge – it has been superbly composed and is a fine example, typical of its period.

Value: Rare £900–£1,500

Scottish Huntsman and Hound
Ecossais Montrant un Renard à un Chien

Mêne

Model circa 1860's
No. 12 in Mêne's catalogue.

Bronze illustrated: Dark, rubbed brass patination – exceptional cast. Height 20 inches approx.

A part of the much larger model already illustrated. Both this and the following groups show Mêne's talent, not only for animal subjects, but for his fine figurative sculpture.

Value: £250–£450
(More for the pair – see next illustration)

Scottish Huntsman with Two Deerhounds
Valet de Chiens Tenant Deux Griffons Ecossais

Mêne

Model circa 1960's
No. 10 in Mêne's catalogue

Bronze illustrated: Dark, rubbed brass patination. Another exceptional cast. A pair to the preceding group. Height 20 inches approx.

Value: Rare £400–£600

Huntsman with Bloodhound *Mêne*
Valet de Limiers

Model exhibited Salon of 1879.
No. 7 in Mêne's catalogue.

*Bronze illustrated: Rich brown patination. A good well-detailed cast. Height 18½ inches.
previous groups. Height 18½ inches.*

In this group, he models perfectly the essential nature of his subject, portraying the concentration of both hound and handler.

Value: £350–£500

Miniatures   **Mêne**

The following are examples of 'miniatures', also very small scale works, which were doubtless a commercial proposition for the artist. On this small scale, whether individual models or evolved by the use of the *appareil réducteur,* they are never as detailed as larger works. They are much sought-after today.

Barbary Stallion (Djinn)

Height 2 inches. Larger model shown earlier.

Value: £150–£200

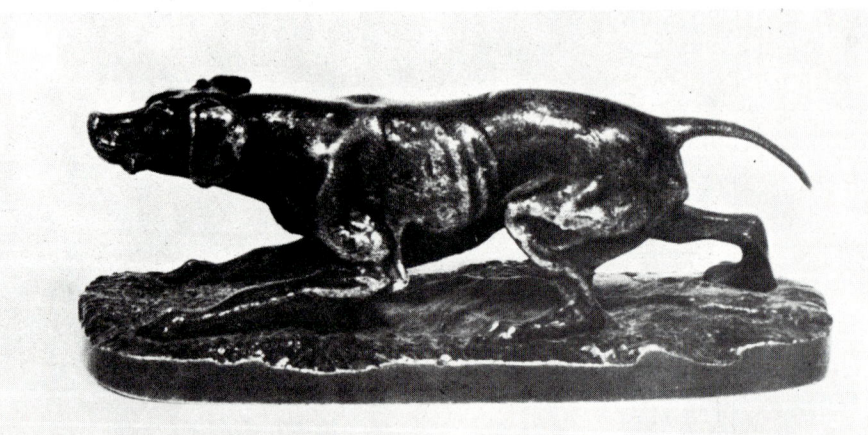

Pointer (Tac)

Height 1¾ inches. Larger model shown earlier.
Value: £60–£85

Miniatures *(continued)*   Mêne

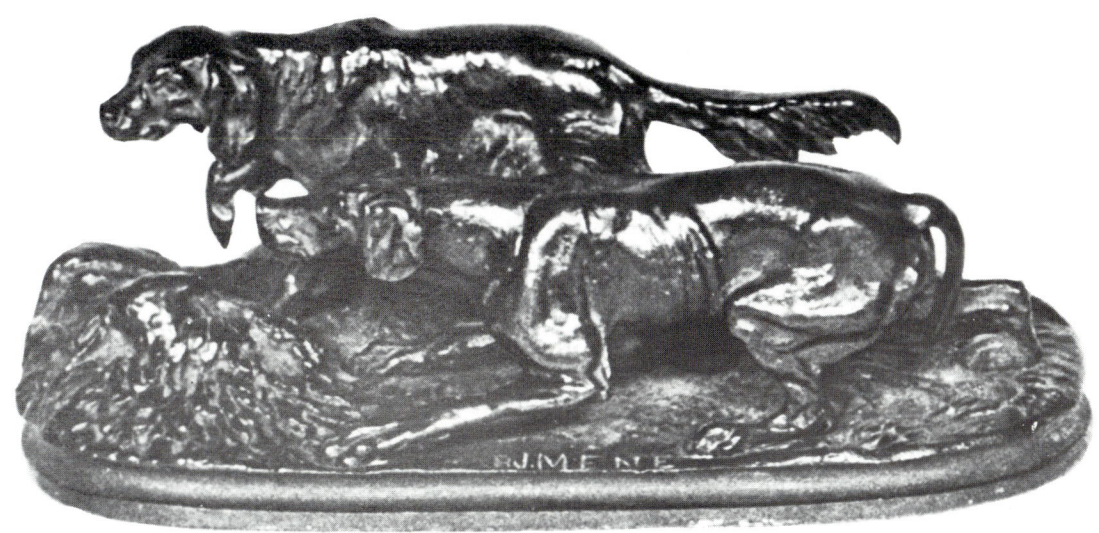

Pointer and Setter

Height 1½ inches. This is the Pointer (Tac) once more, but the setter appears more like the Irish "Médor" than "Sylphie" of the larger model.

Value: £75–£100

Nanny-Goat

Height 1½ inches. Larger model known, but it has a kid suckling.

Value: £50–£75

Mare and Foal

Height 3 inches. No larger model known.

Value: £150–£200

Marie Rosalie (Rosa) BONHEUR 1822-1899

Daughter of the painter, Raymond Bonheur, and sister of Isidore, Rosalie (called Rosa) was born in Bordeaux, but moved with the family to Paris in her early youth. Her father was a landscape artist but, as he was not particularly successful, he decided against an artistic career for his elder daughter. He apprenticed Rosa to a seamstress, but she had neither the temperament nor the aptitude for this work. She eventually persuaded her father into allowing her to study art and attended his drawing classes, as well as studying further with the painter Léon Cogniet.

Rosa Bonheur is included in the animalier school of sculptors on the strength of under a dozen works, mostly modelled prior to 1845, when she won the Gold Medal for painting at the Salon. As well as winning the medal (also a Sèvres vase) she received an important commission from the State, and her celebrated career as a painter began. The admiration for her work and the esteem in which she was held was phenomenal for a woman in the nineteenth century. The 'Landseer' of France, she received decorations and awards, and was on personal terms with the Crown and officials of State; her home was protected by the army during the siege of Paris under orders from Prince Frederick. In England, too, she enjoyed the friendship of Queen Victoria and other noble patrons.

Rosa's success lay in her 'natural' portrayals of animals. Where possible she worked directly from nature, travelling to the Pyrenees and the Scottish Highlands, also spending much time in the country surrounding Paris and in its slaughterhouses. On these local excursions, in order to study undisturbed, she adopted men's clothing – not the 'glamour-girl' variety of her contemporary and friend Georges Sand, the great advocate of female emancipation, but the beret, baggy trousers and blouse of the Breton countryside. In spite of this eccentric garb, which later became her normal wear, and other unconventional conduct, she was a respected and notable person, attracting people of importance, interest and talent through her own admiration for the arts and through her strength of character and personality. Whether her eccentricity was studied or genuine, it did not go unnoticed and may well have been instrumental in her success. She established a career for herself as a professional painter, not a talented amateur, a considerable achievement for a woman at that time. One nevertheless wonders whether an eye-brow would be raised or not today at her absorption with the slaughterhouse; the strange choice of a goat as an apartment pet; or such bizarre behaviour as driving a cab from Fontainebleau station, dressed up as a priest and wearing the ceremonial Legion of Honour sash, whilst inside the vehicle was the renowned English painter, Sir John Millais. Amongst the English, there was the Queen, also Rosa's other noble patrons to be considered, and one imagines that her visits here were a serious matter and that eccentricity was kept in check.

Rosa Bonheur's work was understandably very popular in Victorian England, a country she visited on numerous occasions. In 1853 she went to the Highlands of Scotland, a visit arranged by Monsieur Gambart, the print publisher, who had a good nose for business and wanted paintings *à la* Landseer from Rosa. He was probably well rewarded, as her admiration for Landseer's painting and the Scottish setting generally had a profound influence on her work. Rosa Bonheur certainly did meet

Landseer, but whether on this visit or later is not known. In a letter of 1856 she writes that "he is the greatest painter of animals and I believe that he will remain the greatest of his kind", with more about the poetic grandeur and rare intelligence of Landseer's work. This Rosa — Landseer harmony had a decided influence through Rosa on the animalier school, and on the sculptor Mêne particularly, as well as having an interesting sequel. There was an improbable, but very credulous, rumour circulating that Landseer intended to marry her. When tackled by the painter Frith about it, he declared it a good idea, but that was all. It would certainly have been a splendid commercial partnership, with an output of paintings, one or two of which reached the £10,000 bracket, from two of the most popular artists of the day.

This extraneous detail, gathered from the considerable biographical information available about this artist, would perhaps be irrelevant if a conclusion could not be reached quite simply, that the originality and individuality so strongly marked in Rosa Bonheur as a person has somehow by-passed her art. She was a superb draughtsman, nevertheless, and the worth of her sculpture lies in its basic realism — not 'humanised'; her pictorial portrayal of the animal is absolutely as it exists in nature.

Reclining Bull   Rosa Bonheur

Bronze illustrated: Good rich dark patination, but variable surface detail. *(In a 'static' model of this kind, this is particularly important)*. Height 6½ inches.

One of the artist's favourite subjects. This model shows her keen observation and study of the species.

Value: Rare £150–£250

Reclining Ewe (unshorn), Grazing Ewe (shorn)   Rosa Bonheur

Model circa 1845

Bronzes illustrated: Medium dark patination. Two exceptional casts by Hippolyte Peyrol. Surface detail is superb. Height 4 inches and 5¾ inches respectively.

Outstanding models, they show well the artist's preoccupation with texture and absolute naturalism.

Grazing Ewe — illustrated Antiques International

Value: Reclining Ewe, rare £250–£350
Grazing Ewe £200–£300

Bull   **Rosa Bonheur**

Bronze illustrated: *Rich brown patination – good overall detail. Signed in full, "Rosa Bonheur", and dated 1843. Cast by Hippolyte Peyrol, this was possibly a special commission or actual Salon exhibit. Very unusual size. Height 12½ inches.*

This "beefy" type was probably the result of Rosa's excursions to the slaughterhouse.

Value: Rare £650–£850

Bull   **Rosa Bonheur**

Bronze illustrated: *Gold brown patination — the rubbing that has occurred is not detrimental here, giving light and shade to the cast which is a bit lacking in surface detail. Height 6½ inches.*

A very fine model. The treacherous character is splendidly portrayed.

Value: Rare £250–£450

Bull Rosa Bonheur

Bronze illustrated: Mid-brown patination. Good detailed cast by Hippolyte Peyrol. Height 6 inches.

This model is very typical of this artist's work in sculpture, more 'passive' than much animalier work.

Value: Rare £200–£350

Emmanuel FREMIET 1824 - 1910

Frémiet was born in Paris into an artistic family. His Aunt Sophie, wife of the celebrated sculptor Rude, gave him his first lessons in drawing; later he worked with an uncle, the natural history painter, Werner. Frémiet covered a variety of unpretentious, even macabre, occupations in his pursuit of a career in art before being accepted as a pupil by his uncle, Rude. He executed lithographic drawings, modelled sacred subjects for commercial sale, made anatomical specimens in wax for the Museum of Medicine, and was at one time painter to the Morgue, repairing blemishes on bodies that were to be preserved. Attempting to discourage Frémiet, his uncle Rude had pointed out the hazards and difficulties of a sculptor's career, but without success, and his highest hopes were realised when he was taken into the famous studio at the Rue d'Enfer. Frémiet's determination proved worthwhile. He became one of the most successful and important sculptors in nineteenth century France. He received numerous prizes and awards during the course of his long and successful career. He succeeded Barye as Professor of Drawing at the Museum of Natural History; was made a Grand Officer of the Legion of Honour, and also elected as an Associate of the Royal Academy in England. He was over-loaded with state commissions; it is said that he received many, through the influence of Rude, that should have gone elsewhere, but, be that as it may, he satisfied his many patrons and was a popular and respected figure.

The many monumental works commissioned for Paris and the provinces of France inspired appreciation of Frémiet's work which spread rapidly abroad, and his sculptures were placed in such diverse places as Bucharest, Baltimore and Port Said. Much of this work — pictorial in form and lacking fire — is overburdened with descriptive detail and cannot be classed higher than the mediocre today. There are exceptions, nevertheless. His "Jeanne d'Arc", the gilded bronze equestrian statue which stands in the Place de Rivoli, is a sculpture which captures pride and spirit in both horse and rider; the sculpture of his master, Rude, with its sure and clear cut characterisation of the man, is probably one of Frémiet's finest portrait groups. Although much of his monumental sculpture has been discounted, and even destroyed, there was great prestige attached to it at the time, and Frémiet, a modest man, was nevertheless proud of his success in this field. Particularly emphatic that he was not an animalier sculptor, it is ironic indeed that it is his small sculptures of animals that has brought his name to the fore today.

Almost all the subjects we associate with his animalier work were executed in the early part of his career. He made his début at the Salon in 1843 with a gazelle, in plaster, which was then followed by a wide and varied selection of cats, dogs, horses, bears and many other creatures. All definite individuals, the small dogs and cats particularly are sculpted with such sympathy and tenderness that one imagines that they were made more for the artist's satisfaction than for financial gain, although these charming and lifelike portraits were exhibited with great success by Madame Frémiet in their studio, and later sold from a modest shop. The romantic conception in some of his more ambitious compositions shows itself in the combats of animals and in such vivid imaginative works as — "The Centaur Teree, capturing the Bear", "The Gorilla carrying away a Woman", and the "Fight between a Native and a Bear".

Frémiet's dedication to animals as a subject in his earlier work is apparent, and they are seldom missing from the later, large scale works. Although before his entry into Rude's studio, he had already had a chance to study human anatomy, he had already devoted much time to drawing his animal subjects from life at the Jardin des Plantes — a practise he continued after taking up his career as a sculptor. He was a tireless worker and an ardent researcher, not only in the anatomical field, but also in the numerous varieties of dress he needed for his monumental portraits and equestrian groups. The uniforms, armour, saddlery and harness that Frémiet depicted, were all meticulously correct.

His family originated from the country, and Frémiet had a countryman's love and understanding of animals which is personified in his animalier sculpture. Receiving all the accolades, he no doubt enjoyed his success, but was not spoilt by it, being content to live a simple life surrounded by his family.

Arabian Dromedary
Dromadaire

Frémiet

Model Salon of 1847 (wax)
Listed Frémiet's catalogue of 1859–1860.

*Bronze illustrated: Rich dark brown-black patination. Stamped Frémiet Very good cast.
  Height 11 inches.*

Excellent model — the character of the animal is well portrayed.

Value: Rare £250–£400

Barge Horses   Frémiet
Chevaux de Halage

Model exhibited Exhibition Universelle 1855
Listed Frémiet's 1859–1860 catalogue.

Bronze illustrated: Medium brown patination. The brass base is part of this excellent cast. Height 9 inches.

Expressive composition. Another unusual subject by Frémiet. Also found in Terracotta. Example of this work in possession of the Museum of Mans.

Value: Rare £200–£360

Two Racehorses and Jockeys
Chevaux de Course

Frémiet

Model Salon of 1855. Exposition Universelle 1889.

Bronze illustrated: Rich mid-brown patination – fine quality cast by the founder F. Barbédienne. Height 18 inches.

It is necessary to see this very graceful work from all angles. It is certainly one of the finest models of this subject to be found in animalier sculpture.

Writing in Frémiet's lifetime, Jacques de Biez translates as follows:–

"M. Frémiet recently published with his editors, M.M. Boussod and Valadon, two studies of racing horses, which add the thoroughbred to the series of horses by this artist. A group of jockeys in the saddle and a portrait of the stallion, Barberousse, they are much more documentary than simply ornamental – the historiographer has left his mark of perception and observation."

Value: Rare £1,500–£2,000

Note: Of the three proofs known of this model – the two examined were cast by Barbédienne

Horse   Frémiet

Model circa 1859—1860

Bronze illustrated: Rubbed silver patination — fine cast. Height 11 inches approximately.

In the best tradition of Frémiet's proud and very spirited "charger" horses. Possibly the same model as the "showman's horse" (Salon of 1859).

Illustration: courtesy Sotheby & Co.

Value: Rare £600—£800

Recumbent Husky   Frémiet

Model circa 1860–1880

Bronze illustrated: Light bronze patination. Fair cast, mounted on rouge marble. Height 4 inches.

Good lively animalier model.

Value: £60–£100

Heron   Frémiet

Model circa 1857–1860

Bronze illustrated: Medium dark patination, reasonable cast. Height 7 inches.

This model shows Frémiet's talent for portraying the character of his subject.

Value: £120–£200

Seated Setter   Frémiet

Model circa 1860–1880

Bronze illustrated: Medium dark patination. Good crisp cast. Height 6½ inches.

Sensitive animalier model possibly earlier than stated. (See the various dogs and hounds listed in Frémiet's catalogue of 1859–1860, Appendix D).

Value: £100–£200

Pair of Seated Dachshunds Frémiet
Chiens Bassets

Model circa 1850's.

Bronze illustrated: Rich brown patination – good quality cast. Height 5 inches.

Excellent modelling – a very natural and lifelike composition. "Chiens Bassets" are listed in 1848, and the portrait group "Ravageot et Ravageole, Chiens Bassets" (see Salon list) in the early 1850's. They are most likely the same group, as only one appears in the 1859 catalogue.

Value: £150–£250

Seated Dachshund   **Frémiet**
Basset Seul

Model circa 1860-1880

Bronze illustrated: *Medium dark patination — good quality cast. Height 5½ inches.*

This model is similar to the upright hound in the preceding group. Good realistic work.

Value: £100–£200

Stretching Dog   **Frémiet**

Model circa 1860–1880

Bronze illustrated: Dark brown patination – nice quality cast. Height 3¾ inches.

One of Frémiet's good animalier works.

Illustrated – "Antiques International"..

                             Value: £100–£200

Wounded Dog
Chien Blessé Couché

Frémiet

Model circa 1860

Bronze illustrated: Rich dark patination – good quality cast. Height 3½ inches.

Interesting and unusual subject matter – typical of Frémiet's very individual animalier subjects.

Value: Rare £100–£200

Greyhounds   **Frémiet**

Bronze illustrated: This bronze is the property of the Victoria and Albert Museum. Silvered bronze. Acquired by the Museum in the lifetime of the artist (1896). Height 7½ inches.

Impressive, well modelled group – Rare.

Illustration by courtesy of the Victoria and Albert Museum, London.

Seated Hound Frémiet

Bronze illustrated: This bronze is the property of the Victoria and Albert Museum. Medium brown patination. Height 11 inches.

A most interesting and rare model.

Illustration by courtesy of the Victoria and Albert Museum, London.

Seated Cat   **Frémiet**

Model circa 1850

Bronze illustrated: Silvered bronze. Not an exceptional cast, but rare in this patination Height 3¼ inches.

A very successful, lifelike model.

Example of this work in the possession of:-

        Walters Art Gallery – Baltimore

          Value: £80–£120

Cat and Two Kittens Frémiet

Model circa 1849 (possible Salon exhibit of that year).

Bronze illustrated: Rich dark patination. Fine quality cast. Height 3 inches.

Subject well treated – typical of the artist's animalier work.

Value: Rare £100–£200

Cat with Newborn Kittens
Chatte Mangeant ses Petits

Frémiet

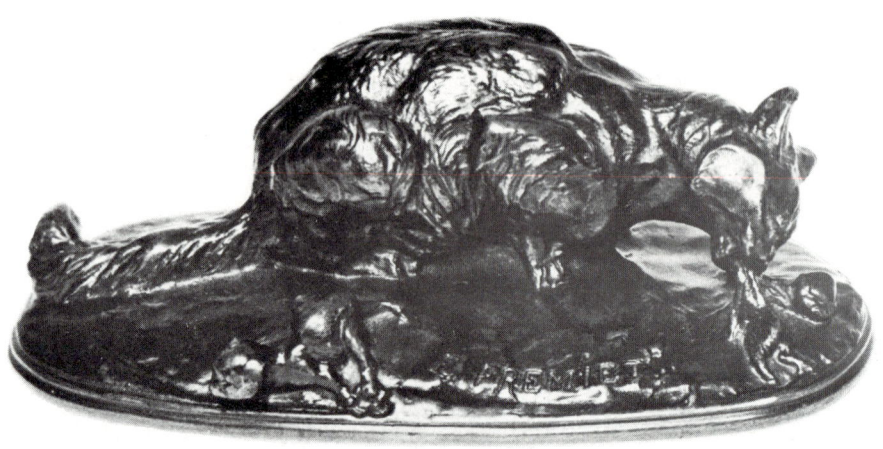

Model circa 1849

Bronze illustrated: Rich dark patination – a fine well detailed small cast. Height 2¾ inches.

A beautifully modelled work of a unique subject. (It is hoped that the cat is removing the 'caul', giving life, not destroying it.) It is a companion piece to the previous group.

Value: Rare £100–£200

Seated Cat **Frémiet**

Bronze illustrated: A finely detailed small work. Height 3 inches.
Illustration by courtesy of the Walters Art Gallery, Baltimore.

Goat and Kid Frémiet

Model circa 1859.

Bronze illustrated: Dark patination. Fair quality cast. Stamped Frémiet. Height 7 inches.

Good naturalistic composition.

Similar cast in the possession of the Victoria & Albert Museum, London, (on view at the present time at their Museum at Bethnal Green).

Value: £100–£200

Centaur Capturing a Bear   **Frémiet**

Plaster model, Salon of 1861

Bronze illustrated: Dark green patination. A fine quality cast. Height 14½ inches.

An impressive model in the romantic vein, its full title is 'The Centaur Teree carrying to his lair a bear taken in the Hemus Mountains'.

Example of this work in possession of:-

    Bonnat Museum – Bayonne

        Value: Rare £450–£650

Napoleon III Mounted on Phillippe — Frémiet

Model 1855–1860

Bronze illustrated: Dark green patination. Good quality cast. Height 15 inches.

The model, executed between 1855 and 1860, was part of a collection representing the various services of the French Army. This collection, commissioned by the Emperor, was destroyed in the fire at the Tuileries. (The model for the statuette of Napoleon III, which was the only one saved, was later in the possession of the Empress Eugénie).

Illustrated – "Antiques International".

Value: Rare £450–£650

Isidore Jules BONHEUR 1827-1901

Isidore Jules Bonheur, son of the painter Raymond Bonheur and brother of the famous Rosa, was born at Bordeaux. He studied painting at first with his father, but later went to the Ecole des Beaux Arts. He made his début at the Salon of 1848 with a picture and a plaster group of the same subject – "An African Horseman attacked by a Lion". He exhibited regularly at the Salon from then on, and also at other exhibitions, winning medals in 1865 and 1869 and also the prized Gold Medal at the Exposition Universelle of 1889. He had given up painting in favour of sculpture early on in his career, and, though noted primarily for his small groups, he did complete two large scale commissions – the monument to his sister, Rosa, at Fontainbleau, and the two stone lions that ornament the steps of the Palais de Justice, Paris.

Many of his bronzes were edited by Hippolyte Peyrol, who was his uncle by marriage. The Peyrol casts for both Rosa and Isidore Bonheur are exceptionally well done, which would suggest a close understanding between founder and sculptor.

Working in the realist manner, Bonheur was not a particularly original sculptor, but there is little doubt that he was an acute observer of nature – his animals were not 'humanised', but modelled to catch a movement or pose characteristic of the particular species. He achieved this completely with sculptures of horses, which, usually more relaxed than spirited, are among his most successful works.

Bull and Cow   **Isidore Bonheur**

Bronze illustrated: Medium brown-black patination. Good detailed cast. Height 12¼ inches. A natural and well composed group – unusual.

Value: Rare £150–£250

Two Bulls   **Isidore Bonheur**

Models plaster. Salon of 1865

Bronzes illustrated: Rich dark patination. These fine casts were sold by Henry Spencer and Sons at a Yorkshire sale for £1,000. Height 15½ inches.

Superb models. Shown at the Salon in plaster, they were ordered for the Sultan's Palace, Constantinople. (Presumably in bronze on a large scale). Whether they were completed and reached their destination, and what has become of them today, is not known.

Example of the right hand model in possession of the Victoria and Albert Museum, London.

Illustration: Courtesy Henry Spencer & Sons, Retford.

Value: Rare £400–£600 each
Pair up to £1,500

Bull   **Isidore Bonheur**

The bronze illustrated is the property of the Victoria and Albert Museum. Rich brown patination. This fine quality cast was acquired by the Museum in the lifetime of the artist (1886).

This model, one of the two illustrated previously, is less rare than its pair.

Illustration by courtesy of the Victoria and Albert Museum, London.

Cow — Isidore Bonheur

Model exhibited Salon of 1872

Bronze illustrated: Rich dark patination – A fine, well detailed, cast founded by Hippolyte Peyrol. Height 5 inches.

Possible made as a pair to the Bull by Rosa Bonheur, page 180. Good interpretation.

Value: Rare £170–£250

Mare and Foal   **Isidore Bonheur**

Model exhibited Salon of 1859 (plaster)

Bronze illustrated: Dark rubbed brass patination. Brass rim to base. Fine cast founded by Hippolyte Peyrol. Height 7 inches.

A very sensitive and natural model — one of the finest by this artist.

Example of this work in the possession of:-

    The Heritage Foundation — Canada.

Value: £900–£1,600

Ecorché Horse   **Isidore Bonheur**

Bronze illustrated: Height 12 inches.

This sculpture was not intended as a decorative work, but for artist's models or veterinary purposes. Plaster copies were cast by the British Museum for use in Art Schools.

Value: Bronze £400–£550

Racehorse and Jockey 　　　　　　　　　　　　　　　　　　　　　　　　　**Isidore Bonheur**

Model circa 1860's.

Bronze illustrated: Dark patination. Good quality cast. Length 25 inches.

A fine model. The relaxed attitudes of both horse and jockey are excellently portrayed.

Illustration – courtesy of Sotheby & Co.

　　　　　　　　　　　Value: £900–£1,400

Horse and Jockey   Isidore Bonheur

Bronze illustrated: Rich dark patination. Good quality bronze cast, on marble and ormolu mount. Overall height 19 inches.

Interesting model, most probably a portrait group.

Illustration — courtesy H. Morton Lee, Esq., Chichester.

Value: Rare £750–£950

The Prince of Wales, later Edward VIIth, on Horseback — Isidore Bonheur

Bronze illustrated: Silvered bronze. Excellent detail. Founded by Hippolyte Peyrol. Height 16 inches.

Excellent portrait group. This model was probably commissioned by an English regiment or perhaps by the Queen herself. (The artist's sister may have obtained this for him as she was in favour at Court and with the nobility).

Value: Rare £450—£850

Mounted Postillion with two Carriage Horses   Isidore Bonheur

Model exhibited Salon of 1866 (bronze).

Bronze illustrated: Medium dark patination. Good surface detail here – slightly obscured as the cast has been neglected. Height 14 inches.

Unusual and lively model.   The heavier type of horse is well shown.

Illustration – courtesy of Sotheby & Co.

Value: Rare £450–£750

Cavalry Horse and Groom Isidore Bonheur

Bronze illustrated: Rich brown patination – not a very well detailed cast. Height 12 inches. A well modelled, lively 'equine' study.

Value: Rare £200–£300

'Royal' Tiger Isidore Bonheur

Model exhibited Salon of 1868 (plaster).

Bronze illustrated: Dark rubbed brass — exceptional cast. The crisp surface detail is easily seen in the illustration. Height 6 inches.

Very powerful model altogether. (Sculptor's licence must be allowed, as purists will notice the incorrect tail).

Value: Rare £280–£400

Gazelles Isidore Bonheur

Model exhibited Salon of 1853.

Bronze illustrated: Medium dark patination – good surface detail on the cast. Height 4½ inches.

The delicate agility of the subjects is well shown. A good model.

Value: Rare £200–£300

Goose Drinking from a Bucket   Isidore Bonheur

Bronze illustrated: Unusual and rare small sculpture by Bonheur, the possession of the Walters Art Gallery. Height 2½ inches.

Illustration – courtesy of the Walters Art Gallery, Baltimore.

Ram and Ewe   **Isidore Bonheur**

Bronze illustrated: Medium dark patination – reasonable quality cast. Height 5 inches.

This 'pictorial' type of work is derivative of Bonheur's sister, Rosa. Although it is adequately modelled, he has done much better with other subject matter.

Value: Rare £150–£250

Bear, Bonasses, Two Pigs Isidore Bonheur

Bronzes illustrated: Three interesting and very unusual groups, (unfortunately not examined). Average height 8½ inches.

Illustrated — courtesy of Sotheby & Co.

Value: Rare. At least £100–£200+ each for good casts.

Jules MOIGNIEZ 1835-1894

Moigniez was born at Senlis sûr l'Oise, but made Paris his home until the last year or two of his life. He studied there under the sculptor Paul Comolera, and exhibited his work for the first time at the Exposition Universelle of 1855. He became a regular contributor to the Salon from 1859, where he received an honourable mention, until his retirement in 1892. Moigniez's work was highly appreciated in France, America and also in England – he received a medal at the Great Exhibition held in London in 1862 – and it is known that over half his work was exported abroad.

The popularity of Moigniez's work in his lifetime is understandable. His compositions were decorative, in no way likely to offend the squeamish. Purely visual as seen in nature, they were easily understood. In some of his sculptures of animals, particularly dogs, he is derivative from Mêne, but the crispness and general excellence of the founding and finishing saves many from mediocrity. The casting, which is invariably excellent, was in the hands of his father, who started a foundry for this purpose in 1857. Originally a metal gilder, he experimented with a wide variety of patinations, including gilding and the many shades of golden bronze that were so successful and are particular to Moigniez's work.

In his premier choice of subject – the sculpture of birds – Moigniez excelled. He developed an individuality in sculptural form reminiscent of the exotic splendour of the subjects created in the paintings of de Hondecoeter, Casteels and Snyders, who undoubtedly inspired his interest. Moigniez's elegant interpretations of the many varieties of waders, game, and other birds have earned him his place as first in this field of animalier sculpture.

Bronzes illustrated: Rich dark patination. A pair of good quality casts. Height 8½ inches and 9½ inches.

Well modelled. More unusual than game bird subjects in animalier sculpture.

Value: £350–£400 pair
Ruff – £150–£220
Reeve – £100–£150

Group of Partridges   Moigniez

Exhibited Salon of 1865 (Silvered bronze)

Bronze illustrated: Rich dark rubbed brass patination. Good quality cast. Height 15 inches.

Typical of much fine animalier work. This is a complicated composition that succeeds and has not been spoilt by unfortunate additions.

Value: £300–£500

Bird Feeding her Fledglings Moigniez

Bronze illustrated: Rubbed brown and dull gold patination. Exceptional chiselling. Height 21 inches.

A fine model amongst Moigniez's wide and varied groups of this subject. Intricate and delicate composition.

Illustration — Courtesy of Mr. & Mrs. B. Glenn Miller.

Value: Rare £250—£450

Turkey and Cock — Moigniez

Bronze illustrated: Rich brown patination — detail not as finely chiselled as on some of Moigniez's sculptures. Height 8½ inches.

Clumsy composition — Farmyard subject, more suitable for painting than sculpture.

Value: £70–£110

Golden Pheasant — **Moigniez**

Bronze illustrated: Medium to dark patination. The surface detail is superlative — An exceptional cast. Height 16½ inches.

Excellent realist model

Value: Rare £200–£350

Two Swallows Moigniez

Bronze illustrated: Pale bronze rubbed brass patination — Exceptional chiselling. Height 7 inches.

One of Moigniez's most successful compositions of small and very active birds.

Value: Rare £150–£300

Polish Bantams **Moigniez**

Bronze illustrated: Medium dark, rubbed brass patination. Fine quality cast. Height 7 inches. A natural, well-composed model.

Value: Rare £100–£200

Sandpiper   **Moigniez**

Bronze illustrated: Rich brown patination – good quality cast. Height 9¾ inches.

This unusual subject is well-modelled in what would appear as an over-intricate composition. It is far better seen in the round.

Value: Rare £150–£220

Bronze illustrated: Medium patination – good chiselling of cast. Height 13 inches.
A reasonably well-modelled bird spoilt by the over-large, bulky base.
Value: £80–£140

Cock Pheasant and Weasel — Moigniez

Model exhibited Salon of 1864 (bronze)

Bronze illustrated: Rich dark patination – Exceptional chiselled detail on pheasant and weasel (Its head can be seen between foliage and bough – bottom right). Very fine cast. Height 20¾ inches.

When compared with the previous model, the superiority of this excellent work is very evident.

Value: Rare £250–£400

Bronze illustrated: Dark patination — Good cast. Height 27 inches.

This model shows the influence of the flamboyant style of the earlier, Flemish painters and it captures superbly the elegant slow motion of this species. Essentially a life-size work, it is interesting to compare it with a similar work, illustrated later, by the artist's master, Comolèra.

Illustration — courtesy of Sotheby & Co.

Value: Rare £300—£500

Bronze illustrated: Brown, rubbed brass patination – Quite good detail. Height 17 inches.
A fine example of Moigniez's baroque style in sculpture.

Value: Rare £250–£450

Exhibited Salon of 1866 (silvered bronze).

Bronze illustrated: Medium dark patination — Good crisp detail, an excellent cast. Height 21¼ inches.

A similar composition to previous group. The artist is modelling a subject in which he excelled, showing his best and most original work. The treatment of the feathers is exceptional in this medium. (The groups on the back are the so-called 'osprey' feathers, purely a milliner's term as they are not present on the eagle of that name.)

Value: Rare £300–£500

Bronze illustrated: Dark patination – A poor cast. Height 12¼ inches.

This model is one of the parent birds from the group already illustrated. The alert, watchful expression looks merely aggressive in the single sculpture.

Value: £80–£150

Sparrows Fighting, Cock and Weasel   **Moigniez**

Model exhibited Salon of 1867 (Sparrows – bronze).

Bronzes illustrated: Two examples of Moigniez's small sculptures of birds. Both are good quality and the difference in the patinas is easily seen. Dark bronze and gilt. Height approx. 4½ inches.

Good models of the artist's typical animalier work with these subjects.

Value: Sparrows Fighting £80–£150
Cock and Weasel £40–£60

Hare   **Moigniez**
"Chasse Ouverte"

Bronze illustrated: Rubbed brass and dark patination. A really fine crisp cast – superb quality. Height 11½ inches.

Good interpretation of the animal – excellent alert expression.

Value: Rare £200–£300

Merino Ram   **Moigniez**

Model exhibited Salon of 1861 (wax)

Bronze illustrated: Dark patination. Fine quality cast (much detail obscured by dust). Height 9½ inches.

Moigniez modelled several large groups of sheep. Overloaded, the subjects fail to capture the presence and bearing apparent in this single example.

Value: Rare £180–£250

Dachshunds Hunting Moigniez

Model circa 1855–1860

Bronze illustrated: Dark patination – chiselling erratic and ill-defined. Height 4 inches.
A lively and lifelike composition.

Value: £100–£200

Pointer and Partridge, Setter and Rabbit Moigniez

Pointer exhibited in plaster at the Exposition Universelle 1855. Bronze, Salon of 1859.

Bronzes illustrated: Brown brass patination – Not the best quality cast. Height 7¾ inches.

Larger originally and bought by the State, the Pointer model was popular and cast in small sizes. Moigniez modelled the Setter as a companion piece. Both are derivative from Mêne.

Value: £70–£140 each

Setter and Pointer — Moigniez

Bronzes illustrated: Medium dark patination — A good quality pair. Height 9 inches.

There is more 'life' in these two models, although they are more statically posed, than those in the previous illustration.

Illustration — author's Collection.

Value: Rare £100–£170 each

King Charles Spaniel   **Moigniez**

Model exhibited Salon of 1861

Bronze illustrated: Brown rubbed brass patination – Exceptional crisp detail. A very fine cast. Height 14½ inches.

Excellent realistic model.

Value: Rare £500–£750

Wolfhound — Moigniez

Model circa 1861

Bronze illustrated: Dark patination – surface detail could be crisper. Height 9¾ inches.
The hound is well modelled – the tortoise is a bit out of place.

Value: Rare £70–£150

Irish Setter **Moigniez**

Model circa 1861

Bronze illustrated: Brass-brown patination. A very fine cast altogether. Height 8¾ inches.

This is probably the Scottish hound setting. It is one of the artist's best models of this subject.

Value: Rare £220–£350

Stallion
"Chief Baron"

Moigniez

Bronze illustrated: Mid-brown patination. Good cast. (Intricate detail here would be superfluous and was not intended). Height 11 inches approx.

Moigniez has modelled this portrait in the 'romantic' style. (It may be that it was impossible to portray this horse other than in movement; he was reputed to have been a handful, needing five grooms to un-box him on his arrival in England from France). A very pleasing, graceful work.

Value: £550–£750

Mare and Stallion — Moigniez

Bronze illustrated: Rich brown patination with plenty of light and shade. Some of the detail is good, but overall, the cast could be a little sharper. Height 13 inches.

When compared to a similar model by Mêne, this may appear to lack 'punch', but the placing of the pair is excellent. A natural, well-composed group.

Value: Rare £650–£900

Thoroughbred Mare
"Mon Etoile"   **Moigniez**

Bronze illustrated: Rich dark brown patination. Excellent cast. Height 13¾ inches. A very fine portrait model.

Value: Rare £550–£850

Arab Stallion **Moigniez**

Bronze illustrated: Rich mid-brown patination. Good detailed cast. Height 10 inches. One of the models from the artist's "Mare and Stallion" group.
Value: £450—£600

Bronze illustrated: Entitled "Avant la Course", in addition to the signature on the base, this bronze also bears a brass plaque inscribed "Société hippique d'Oran Prix Gentlemen 1894". Rich brown patination. Fine detail cast. Height 10 inches approx.

In the realist manner, this model was certainly executed earlier than the date of this 'gentlemen's' prize. Moigniez had ceased working before 1894, the year of his death. There is an eighteenth century flavour about the jockey, who is somewhat over-horsed.

Value: Rare £500–£700

The Kill. (A Boar Hunt)   **Moigniez**

Bronze illustrated: Rich dark patination.   *Good cast (much detail obscured by dust).*
Height 23 inches.

Very much in the nineteenth century French taste, this large impressive model is more for the board-room than the private collector today. The whole group suggests pace — the desperate, fleeing boar, the determined hounds, also the concentrated and very 'professional' look on the hunter's face, are exceptionally well-modelled. Compare this to Mêne's fine, but altogether more peaceful sculpture — "Hunt in Scotland".

Value: Rare £500–£750

Pointer and Partridge Moigniez

Bronze illustrated: A smaller edition of a subject illustrated previously, this cast is good quality for groups of this size. Height 4½ inches approx.

These small bronzes are much collected, but are really only of interest because of the appeal of their size.

Value: £25–£50

Bull and Cow Moigniez

Bronze illustrated: A miniature of a larger original. It has lost much detail in the reduction (presumably with the aid of M. Collas's machine). The bronze has suffered some damage, which is invariably the case with examples as small as this. Height 2¼ inches, excluding base.

Value: £20–£45

Casket, ornamented with bird subjects Moigniez

Bronze illustrated: Silvered bronze. A very fine and detailed work. Height 4½ inches approx.

Moigniez produced several of these small caskets, showing one at the Great Exhibition of 1862 in London. Usually velvet lined, they were probably intended as jewel boxes. In the 'baroque' style, this work is a singularly attractive *objet d'art*.

Illustration – Courtesy of H. Morton Lee Esq. Chichester.

Value: Rare £80–£160

Additional Artists with work illustrated

Alphonse-Alexander ARSON   1822–1880

Arson was a pupil of Combette. He specialised in the sculpture of birds, both domestic and wild, and exhibited regularly at the Salon from 1859 until his death. A silvered bronze group of "Pheasant and Chicks" is listed as Arson's exhibit in the Salon of 1870.

Alfred BARYE   19th Century

Son of Antoine-Louis Barye, Alfred studied with his father. He exhibited a number of sculptures of race-horses at the Salon from 1864 to 1866. He exhibited spasmodically, and later entries include a "Group of Partridges" in 1874, and a figurative work "A Sixteenth Century Buffoon" in 1882. A competent sculptor only, without the original talent of his celebrated father, he produced some adequate animalier work. His portrait groups of horses are among his more interesting works.

Auguste Nicholas CAIN   1821–1894

Cain was a pupil of Rude. He married Mêne's daughter in 1852 and was much influenced and helped by this artist. He shared his father-in-law's foundry, but unlike Mêne, he accepted many government commissions for works destined for important buildings and public places. He made his début at the Salon of 1846, and in the period between 1847 and 1888 thirty-eight works are listed as commissioned or exhibited at the Salon. In 1879 he created the colossal equestrian statue of Herzog Karl von Braunschweig for Geneva. Adjacent are the two huge lions and two griffins, all in red marble.

He edited his own bronzes in Mêne's foundry, which he continued to run after the latter's death, although some of his very large sculptures were cast by F. Barbédienne.

Working for almost all his life with his father-in-law, many of Cain's small compositions are derivative of Mêne. The Susse Frères found him a 'commercial propo-' sition' nevertheless, buying some of his models and issuing a catalogue of his work shortly after his death.

His larger works depicting life and death struggles of animals are by far his most successful; the "Tiger and Crocodile" in the Tuileries gardens is magnificent.

Paul COMOLERA   1818–1897

Comolèra was a pupil of Rude. After completing his studies at the Rue d'Enfer he made his début at the Salon of 1846 with a group of "Golden Pheasants of China". He exhibited regularly until his death. He was known chiefly for his sculpture of animals and birds, and was the master of Moigniez, who was his most outstanding pupil.

Some of Comolèra's bronze sculptures were reproduced in fayence at Choisy by H. Boulanger & Company.

Comolèra's work was edited by A. Gouge.

Paul-Edouard DELABRIERRE   1829–1912

Delabrierre is best known for his small groups of animals, many with figures, in bronze, terra cotta and plaster. He was a pupil of the painter Delestre, and exhibited regularly at the Salon from 1848 to 1882. In 1857, he executed a large composition for the façade of the Louvre entitled "The Art of Riding", and there is also his "Indian Panther and Heron" in the Museum of Amiens.

Hunting scenes and animal subjects predominate in Delabrierre's work, and the following are among his most successful: "Stags Fighting", "Family of Dogs", "Lion and Crocodile". "Rider attacked by a Tiger", "Fox and Wild Duck", "Arabs Hunting with Birds", "Picador".

Alfred DUBUCAND   B. 1828

Dubucand was a pupil of Lequien. He exhibited at the Salon from 1867 until 1883, receiving a medal in 1879. Among his work the following are listed. "Huntsman Restraining Hounds", "Griffon Attacking a Duck", "Returning from the Hunt" and "Ostrich Hunt in the Sahara".

Georges GARDET   1863–1939

Georges Gardet, son of the sculptor Joseph Gardet, was a pupil at the Ecole des Beaux Arts under Aimé Millet, and also of Frémiet. He showed a talent far superior to many of his contemporaries and became an extremely successful sculptor, working in marble and bronze. Many replicas of his work were produced in Sèvres porcelain. He achieved his first important success at the Salon of 1887 with a group, "Panther and Python". He exhibited regularly at the Salon from 1883 and also worked abroad. He was made a knight of the Legion of Honour, also a member of the Society of French Artists and of the Academy of Fine Arts. He received many medals and awards, winning the Grand Prix at the Exposition Universelle in 1900.

His important groups are the following: "Fighting Panthers", "Mouse" (Sèvres porcelain, Luxembourg Museum, Paris), "Two Panthers" (Simu Museum, Bucharest), "Lion and Lioness", "The Dogs of Chantilly" (Limoges), "Panthers", "Parrots" (Paris Museum of Modern Art), "Great Dane" (Petit Palais), "Bison attacked by a Jaguar" (Laval) and "Fallow Deer" (Porte Dauphine, Paris). A sculpture of panthers by Gardet was purchased by the Ashmolean Museum, Oxford in 1964.

Joseph Raymond Paul GAYRARD   1807–1855

Gayrard was a pupil of Rude and David d'Angers. He exhibited at the Salon from 1831 to 1855 and was very much appreciated in high society, executing a number of sculptures of celebrated people. He received a second class medal in 1834 and first class in 1846. Of particular interest to us is his beautiful harness horse entitled "Cheval d'Attelage, harnaché et bridé" which was accepted as a plaster in the Salon of 1847 and was acquired by the Commissioners of the Musée de Valenciennes in 1850. This model was reproduced in bronze and exhibited at the Salon in 1848. It would appear that Gayrard did not make a speciality of animal subjects. The three illustrated are, indeed, rare, as almost all his works listed – either Salon exhibits or commissions – are sculptures of the human form. Examples of this work is in museums at Caen, Le Havre, Rodez, Tours and at the Comédie Française theatre in Paris.

John Willis GOOD   19th Century

One of the rare English animal sculptors. All that is known is that Willis Good exhibited at the Royal Academy between 1870 and 1878. His sculpture, entirely of horses, is usually well cast. Sometimes it was founded by Elkington's and bears their stamp.

Pierre LENORDEZ   19th Century

Lenordez exhibited regularly at the Salon from 1855 to 1877. There is an example of his work in plaster in the Museum of Avranches "Captain Estonville at the Defence of Mont Saint Michel", but horses are his subject; in his animalier sculpture Arab stallions, usually named, and mares and their foals, predominate.

Works listed are as follows:-
The Baron, Stallion from the Bois de Boulogne Stud. Wax. Exposition Universelle 1855.
Stallion and Brood Mare – Plaster, Salon 1851.
Part-bred Mare – Plaster, Salon 1867.
"Capitulation of Sedan" – Terra-cotta, Salon 1874.

Arthur-Marie-Gabriel, Le Comte du PASSAGE   1838–1909

Du Passage, although a nobleman and destined for a military career, studied sculpture with Barye and P.J. Mêne, making his début at the Salon of 1865, with a sculpture of a hare. He created small groups of hunting subjects and other animals, principally horses. His sculptures are comparatively rare, but examples of his work can be found in the Museums of Amiens and Orleans.

Ferdinand PAUTROT   19th Century

Pautrot was born at Poitiers in the first half of the nineteenth century. He was almost entirely a sculptor of animals and birds, exhibiting his work at the Salon from 1861 to 1870.

Le Comte G. de RUILLE   19th Century

Details of this French artist are not known. Examples of his work are rare and of fine quality.

Emmanuel de SANTA-COLOMA   19th Century

He was born in Bordeaux. He made his début at the Salon of 1863 and was a contributor until 1870. His date of birth is unknown, his death is recorded as about 1886.

Charles VALTON   1851–1918

Valton was a pupil of Barye, also of Frémiet. He made his début at the Salon in 1868, and in 1875 he became a member of the Society of French Artists. He received medals for his sculptures in 1875 and 1885, and also at the Exposition Universelle of 1889 and 1900. Valton's works in museums include his "Mammoth and Polar Bear" in the Natural History Museum, Paris, "Head of a Lion", at Castres, "Lion", at Constantine, and "Tiger and Tigress" at the Musée Galliera, Paris.

After his début at the Salon he exhibited a further seventy works there in bronze and bronze and marble; also a marble and ivory Boar.

Pheasant and Eight Chicks   **Arson**

Bronze illustrated: Silvered bronze – very fine quality. Signed ARSON, inscribed "Admis aux Beaux Arts 1864". Height 25 inches.

Works by this artist are rare. There is a variation of this group, a "Pheasant and Five Chicks", and his Salon exhibit of 1870 lists an identical subject, but does not give measurements or number of young.

Value: Rare £400–£500

"Vermout". Thoroughbred Horse **Alfred Barye**

Bronze illustrated: Medium dark patination. Signed A. BARYE FILS*. Dated 1864. Additional details on brass plaque. Height 10 inches.

Probably a special commission. One of the artist's better works.

Illustration – Courtesy of Sotheby & Co.

                      Value: Rare £500–£900

*Usual signature for Alfred, son of Antoine-Louis BARYE.

Donkey
Ane d'Afrique

Cain

Bronze illustrated: Medium brown patination — good quality cast. As well as signing his name, the sculptor has 'branded' the model with his initials on the flank. Height 5¾ inches.

Unusual subject — very well done. There is a variation of this model with panniers across the back.

Value: Rare £100–£200

Heron
Héron

Cain

Bronze illustrated: Medium dark patination overall, with brass evident. The base, which is part of the sculpture, is brass – a good cast. Length 14 inches.

A fine, original model.

Value: Rare £100–£200

Crowing Cock **Cain**
Coq sur Panier

Bronze illustrated: Rich dark patination – brass base part of the sculpture. Good detail. Height 6¾ inches.

This lively subject was a favourite with Cain. Listed in his catalogue of 1857.

Value: £50–£100

Swallow in Flight   **Comolèra**

Bronze illustrated: *Medium brown patination. Not an exceptional cast, rather lacking in detail. Height 7 inches.*

Excellent subject.

Value: £100–£250

Fighting Cocks — Comolèra

Bronze illustrated: Rich dark patination — black overlaying brown. Very good and detailed cast. Height 17½ inches.

Re-edited, the "Triumphant Cock" appears as a single sculpture.

Value: Rare £300—£400 group
Single work about £200

Dead Bird   **Comolèra**

Bronze illustrated: Gilt bronze on black and yellow veined marble base. Good cast. Height 3¼ inches.

Good model, but not a popular subject.

Value: Rare £50–£70

Deerhound   **Comolèra**

Bronze illustrated: Medium dark patination. Exceptional bronze. Detail very crisp. Height 8 inches.

A fine model – perhaps a portrait commission.

Value: Rare £300–£400

Tiger Delabrierre

Bronze illustrated: Green patination. Mediocre cast. Height 4½ inches.

Many similar subjects by this artist are derivative of Barye and vary considerably in the quality of the castings. His groups in the 'realist' form are much more successful and are usually far better casts.

Value: £50—£75

Two Bears Sparring Delabrierre

Bronze illustrated: Dark patination. A jelly mould construction cast. Lively group but detail could be better. Height 7 inches.

Well-composed model.

Value: Rare £100—£150

Gamekeeper and his Dog.   **Dubucand**

Bloodhound

Bronze illustrated: Rich dark patination. Good cast.

Rather a stilted model.

Value: £80–£110

Bronze illustrated: Medium brown patination. Very crisp and detailed cast. Note fine chiselling on breeches of figure. Height 7½ inches.

Very lively small hounds similar to present day beagles were a favourite subject of this artist. The casting of his work varies considerably.

Value: Rare £150–£200

Tiger Attacking a Tortoise **Gardet**

Bronze illustrated: Rich mid-brown patination. Good cast. Height 7½ inches.
Barye's influence is shown in this model. It is an impressive powerful work.
Value: Rare £300−500

Buffalo Attacked by a Puma Gardet

Bronze illustrated: Rich dark patination. Black over green and brown. Very fine cast. Height 8½ inches.

One of Gardet's most interesting and impressive models.

Illustrated 'Antiques International'.

Value: Rare £350–£550

Borzoi and Whippet   **Gardet**

Bronze illustrated: Brown patination. Good cast by F. Barbédienne. Height 11¼ inches. Elegant group in the realist manner – probably an earlier work by the artist.
Value: Rare £150–£350

Bronze illustrated: Dark green patination. Mediocre cast. Height 7 inches.
Interesting model – impressionist in form.
Value: Rare £100–£150

Harness Horse   **Gayrard**
Cheval d'Attelage Harnaché et Bridé

Original model exhibited Salon of 1847 – plaster. 1848 – bronze.

Bronze illustrated: Rich medium dark patination. Exceptional quality bronze. Height 13 inches.

Very unusual and rare subject.

Example of this work in possession of:-
    Musée de Valenciennes

Value: Rare £800–£1000

"The Monkey Horserace" — Gayrard

Bronze illustrated: Light brown patination. Good crisp detail. Height 7 inches. Inscribed "Paul Gayrard 1846".

Very original subject.

Value: Rare £400—£600

Deerhound   **Gayrard**

Bronze illustrated: Fire gilt bronze. Dated 1848. Exceptional detail. Height 3½ inches. Fine model.

Value: £350–£550

Arab Stallion   **Lenordez**

Bronze illustrated: Rich dark patination. Scroll cast with the base inscribed with name of subject. Height 10 inches approx.

Lenordez did several equine portraits — the smooth, rather stylised treatment is typical of his work in this field.

Value: Rare £400–£600

Welsh Mountain Pony Lenordez

Bronze illustrated: Medium dark patination. Good detail on the cast. Height 8 inches. Very unusual subject.

Value: Rare £250–£450

Setter   **du Passage**

Bronze illustrated: *Variegated patination of brown shades with brass evident. Detailed, good quality cast. Signed "Ct. du Passage" (usual signature for this artist). Height 15 inches.*

Well-modelled subject.

Value: Rare £250–£360

Horse and Groom   du Passage

Bronze illustrated: Medium brown patination. Good cast — groom is particularly detailed. Height 17 inches.

This 'trotter' or 'pacer' is a rare subject, and the model was obviously successful. Cast in three scales, this is the medium size.

Value: £400—£800

Seated Pointer and Setter — Pautrot

Bronzes illustrated: Rich brown patination. Good crisp casts. Height 18 inches.

Exhibited at various exhibitions, this pair are in the possession of the author. The same models (life-size) were bought for £1,800, when auctioned by Messrs. Phillips, Son & Neale in 1968.

*Value: The larger sized groups are rare.
Smaller size (about 9 inches) would be £350
—£400 the pair*

Stag   Pautrot

Bronze illustrated: Dark patination. Good quality cast – inscribed "Admis aux Beaux Arts". (The antlers have suffered some damage and should be in a more upright position). Height 7¾ inches.

Typical animalier model.

Value: Rare £80–£150

Partridge **Pautrot**

Bronze illustrated: Brown patination. Cast of the finest quality, delicate 'feathering' and base detail is superlative. Height 14½ inches.

Good model.

Value: Rare £150–£250

Partridge with Chicks Pautrot

Bronze illustrated: Medium dark patination. Very finely detailed cast. Height 10½ inches. Attractive model.

Value: Rare £150–£250

Macaw   **Pautrot**

Bronze illustrated: Gold bronze patination. Very fine cast. Height 14 inches.

An excellent model by this artist, who is equal to Moigniez in his sculpture of birds.

   Value: Rare £250–£350

Finch   **Pautrot**

Bronze illustrated: Medium brown patination. Good detailed cast. Height 7 inches.

Attractive model.

   Value: £90–£140

Wolf in a Snare   **Pautrot**

Bronze illustrated: Brass-brown patination. Very fine quality cast. Height 7 inches.
A fine model, but not the most popular of subjects.
                       Value: Rare around £100

Horses and Jockeys   **Le Comte G. de Ruille**

Bronze illustrated: This, and one other, are the only examples seen to date of this artist's work.

Illustration by courtesy of Christie's.

Value: £850–£1,250

Horse and Rider **Santa-Coloma**

Bronze illustrated: Dark, rubbed brass patination. Good quality cast. Height 12 inches. Interesting and unusual model. Work by this artist is rare.
Value: £450–£550

Pointing Griffon "Marco"   **Valton**

Bronze illustrated: Brass bronze patination. The superb crispness of detail on this cast is evident. Height 13 inches.

Excellent animalier work.

Value: Rare £300–£500

Two Hounds   **Valton**

Bronze illustrated: *Medium dark patination. Rubbed brass evident where handling has occurred. Height 10¾ inches.*

Well-composed group.

Value: Rare £150–£250

Mastiff — Valton

Bronze illustrated: Mid brown patination. Very detailed chiselling on the cast is evident. Height 10 inches.

Very fine, lively model.

Value: £200–£350

Horse and Jockey — Willis Good

Bronze illustrated: Silvered bronze dated 1875 mounted on pale rouge marble. A fine cast of exceptional quality. Rare in this patination. Height 10 inches.

Fine composition. Excellent subject matter.

Value: Rare £550–£750

Horse and Jockey   **Willis Good**

Bronze illustrated: Medium dark patination. A good quality bronze. Height 11½ inches. One of five known racing subjects executed by this artist — good subject matter.
Illustration — courtesy of Sotheby & Co.
           Value: Rare £400–£600

Pair of Hunting Subjects Willis Good

Bronze illustrated: Rich medium dark patination. Right hand cast dated 1874. Height 11½ inches.

There is a third group by the artist, similar to these, which has the addition of two hounds.

Value: £600–£800 pair

Bronze illustrated: Rich medium dark patination. Height 15 inches. Well modelled, very natural work.

Value: £250–£350

Pair of Racehorses and Jockeys French, last half of 19th century

Bronze illustrated: Light patination. Nice quality casts. Signed W. ROCHE. Height 9½ inches.

Very lively models and excellent subject matter. No details of the sculptor available.

Illustration — courtesy of Sotheby & Co.

Value: £400–£500 each

Deerhound   French, late 19th century

Model illustrated: Dark patination. Well detailed fine cast. Signed DEBUIT. Height 12 inches.

Interesting and unusual bronze. No details of the sculptor are available.

Value: £200–£300

ADDITIONAL SCULPTORS WHO ARE KNOWN TO HAVE MODELLED ANIMALS

Jean-François-Théodore GECHTER   1796–1844

Born in the same year as Barye, he also had the same masters, Bosio and Gros. He exhibited first at the Salon of 1837. He had reasonable success as a sculptor and, although not an animalier, he executed one or two pleasant groups in this vein.

Jean-Léon GEROME   1824–1904

Gérôme was a pupil of Paul Delaroche and is chiefly known as an historical and genre painter. A sculptor, too, he executed much work in this medium, including animals. He made his début at the Salon of 1847 with a group of fighting cocks. He had an important career, receiving awards, decorations, appointments and commissions, as well as finding time to travel widely.

Emile Joseph Alexandre GOUJET   19th Century

A pupil of Barye, he exhibited three works at the Salon between 1868 and 1870.

Hippolyte HEIZLER   1828–1871

He made his Salon début in 1846 and received an honourable mention in 1852. He collaborated on the decoration of the Louvre, the Tuileries and l'Opéra.

Louis-Théophile HINGRE   1855–1924

A pupil of Fervais and Passot, he made his début at the Salon of 1881, obtaining distinctions in 1891 and 1902. At the Expositions Universelles of 1889 and 1900 he received both bronze and silver medals. He was a member of the Society of French Artists from 1909.

Henri-Alfred-Marie JACQUEMART   1824–1896

Born in Paris, Jacquemart entered the School of Fine Art in September 1845, also frequenting the studios of P.Delaroche and of Klagmann. Exhibiting at the Salon from 1847 to 1879, he obtained medals in 1857 and 1865. He was made a Companion of the Legion of Honour in 1870.

His works can be seen in museums throughout France, and much was commissioned for prominent places in Paris and elsewhere. Jacquemart sculpted the large rhinoceros, one of the animals decorating the fountain in front of the Trocadero – the others being the work of Frémiet, Cain and Rouillard respectively – and among his other important sculptures are the four huge lions for the Kasr-el-Nil Bridge, Cairo. Jacquemart's works are also in the following museums:– "Antelope and Snake" (Aix-les-Bains), "Lioness" (Bergues), "Stag" (Chambéry), "Camel" (Nantes).

Dominique LAQUIS   b.1816

Several works are listed as exhibited at the Salon by this artist between 1852–1868.

Lambert Alexandre LEONARD   b.1831

A pupil of Jacquot, Rouillard and Barye, he exhibited animal sculpture at the Salon from 1851 to 1873.

Clovis-Edmond MASSON   1838–1913

Masson was a pupil of Barye and Rouillard. He exhibited many works at the Salon from 1867 onwards, receiving an honourable mention in 1890. The museum at Château-Thierry, also the Walters Art Gallery, Baltimore, have examples of his work.

Georges MALISSARD   b.1877

Malissard was known chiefly for his sculpture of horses. He did some good animalier work early in his life and some fine equestrian groups later, including Albert I, Alphonse XIII, and the Field-Marshals Pétain, Lyautey and Foch.

Victor PETER   1840–1918

Peter was a pupil of Cornu and Devaulx. He made his début at the Salon of 1868 exhibiting with reasonable success until 1900 when he was awarded the gold medal. He was made a Companion of the Legion of Honour in the same year.

Some of his work was cast by the founder Hébrard.

Pierre Louis ROUILLARD   1820–1881

A pupil of Carot, he made his début at the Salon of 1837. Rouillard was commissioned, with Frémiet, Cain and Jacquemart, to execute one of the four groups for the Trocadero fountain. He was made a *Chevalier* of the Legion of Honour in 1886. The museum at La Rochelle has an example of his work.

John Macallan SWAN   1847–1910

A painter and sculptor, Swan began his studies in England, and later went to Paris. He studied painting under Gérome and sculpture with Frémiet. He exhibited at the Royal Academy, London, in 1878. At the Paris Exposition Universelle of 1900 he received the silver medal for painting and the gold medal for sculpture. He was elected an A.R.A. in 1894. Examples of his work are in the Tate Gallery, the Victoria and Albert, and other important museums throughout the world.

WAAGAN   19th Century

Very little is recorded except that Waagan was working in the 1860's. The Sheffield Museum have examples of his animal sculpture.

UNSIGNED BRONZES

The following sculptures, although unsigned, are illustrated for their exceptional quality and for their rarity – with one exception, they are the only examples known of each work.

Toy Dog   French, early 19th century

Bronze illustrated: Light patination bronze mounted on variegated marble. Exceptional detail. Unsigned. Height 3 inches.

A fine small sculpture, rare and unusual.

Value: £150–£200

Greyhound with Dead Hare **French, 19th century**

Bronze illustrated: Medium dark patination. Fine detail. Unsigned. Height 6 inches.
Value: £150–£200

Cavalier King Charles Spaniel **French, late 19th century**

Bronze illustrated: Rich brown patination. Unsigned. Height 7 inches.
Value: £100–£150

Greyhound and Cat 19th century

Bronze illustrated: Silvered bronze, mounted black marble. Good detail. Unsigned. Height 7½ inches.

This bronze attributed to WAAGAN who executed an identical subject.

Value: £150–£200

Horse and Jockey French, 19th century

Bronze illustrated: Rich black patinated bronze, mounted on ochre marble base. Excellent detail. Unsigned. Brass plate bearing inscription:— "This bronze group is represented in the portrait of the late Sir Joseph Duveen by Emil Fuchs now in the Turner Wing of the Tate Gallery". Height 18 inches including base.

Value: £600–£800

One other example only seen of this model. Also unsigned, it was not the quality of that shown above, having dull, brassy patination.

Snail   French, 19th century

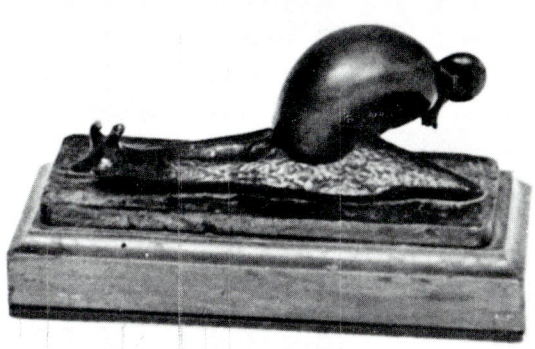

Bronze illustrated: Light bronze patination. Excellent detail. Mounted rouge marble. Unsigned. Height 1¼ inches.

Subject matter unique. Perhaps an early work by Barye.

Value: About £50

CONTEMPORARY PORCELAIN EXAMPLES OF ANIMALIER SCULPTURE

Three Terriers Burrowing

Illustration: *A model in unglazed biscuit or parian porcelain by Copeland & Garrett, after a group by P.J. Mêne. Not the best example; not as well detailed as some.*

Several of P.J. Mêne's sculptures were reproduced in porcelain by this factory at the end of the 19th century. These unglazed models are found comparatively often, but glazed, painted, and decorated ones are rare.

Cock and Hen

Illustration: Fayence. Decorated in brilliant colours and highly glazed, after models by Paul Comolèra.

Several animalier sculptors executed models for the porcelain companies.

Value: £200–£300 the pair

Greyhound

Illustration: *Spelter or Britannia metal example. Unsigned. Late 19th century — Continental. Height 5 inches.*

See notes on founding.

Notes on 19th Century Founding Practice

The *cire perdue* or lost wax process, or that employing the use of sand, were the methods used for casting animalier sculpture (for further details see Appendix A.) Although both these methods required much technical skill and knowledge, the use of sand was proved to be more economic at that time and was in greater use.

In the early part of the nineteenth century the small bronze was once again extremely popular, whether it was reproduced from the 'antique' or a new work. In Paris a whole industry evolved, embracing many hundreds of workers skilled in all the aspects of bronze casting, similar to the rapid growth of the brass founders in England. These craftsmen were instrumental in the success of the animalier school by assisting the craftsmen-sculptors in their own workshops, or by working in the foundries, both large and small, which were at the service of those artists unskilled in foundry practice.

The bronze used for animalier sculpture is almost always that known as 'yellow bronze'. Bronze itself is very often loosely categorised as being a mixture of copper and tin, whilst brass is the same, but with the addition of zinc. This is a very loose description, as bronze invariably contains additions to these basics, either present naturally with the original mined copper, or added by the founder according to his own particular preference. Most animalier bronze when scratched, or with the 'rubbing' which occurs during the course of time, will show a brass base, i.e., yellow bronze, comprised of copper, tin and antimony. There are some with a more copper tinge, but this use of tin and antimony in greater proportion to copper gave a harder surface, thereby ensuring crisper detail to the finished cast, and is more usual.

Patination, or the colouring of the work to give the 'bronze' finish, was done after the casting, either by the use of chemicals or by more unusual means — smoking over fires burning such strange ingredients as willow twigs or old shoe leather. The patinating is an extremely important part of the work, and the skill in producing the many variations is an art on its own. Barye particularly, spent much time developing a variety of colours from the palest to the darkest green and the many shades of black and brown, sometimes underlying one with another, which were such a distinct part of his finished work. Mêne's patinas, too, were rich in tone, although he did not experiment, as Barye did, with the green shades of 'verdigris' to simulate the 'weathered' patinas of early sculpture, but confined himself to the many shades of brown and near blacks particularly associated with small bronzes of the past. Mêne occasionally made use of gilding, or he might leave a cast as it came from the furnace, i.e., firegilt. This was possible only if it showed no patches or streaks caused by the imperfect mixing of the amalgam. Gilded, also gold bronze, patinations were particularly favoured by Moigniez for his sculptures. This may be because his father, who founded much of his work, was a metal gilder. In the main, the use of silver and gilt for patinas was confined to the more important pieces — the special commissions or those for exhibition.

'Lost wax' casts are heavy as a rule, having thick shells, and if there has not been too much cleaning up inside the cast, there is evidence of the core deposit adhering to the inner surface also, rods about one-eighth of an inch thick and in varying lengths which supported the structure before the pour, are occasionally left in. The surface of the bronze has a waxy look and the detail defined without the use of chisels. In sand casts, if, again, they have not been tidied up too much inside (and only if they are the hollow 'jelly mould' form without the addition of a base), one can find one or two of the small pins, about one-sixteenth of an inch thick and under half an inch in length, which held the two moulds together. Sometimes the remains of a dark grey sand deposit can be seen, also the skilful 'jointing' of the parts, but it must be appreciated that a sculpture that is fixed on to a separate base is impossible to examine in this way. It is important to study the outer face of a sand cast bronze for the detail put in by chiselling, (e.g. Barye would vary the scaling on a snake, even omitting it altogether in places, to give the artistic effect he wanted, whereas on an artisan or commercial foundry finished cast there might be an overall and regular pattern on the reptile which is boring and uninspiring). More intricate altogether than the smooth surface of Barye — who relied on his patinas for depth and texture — are the forms created by the 'realist' animaliers. This surface detail is very necessary to add 'finish' to their work; one should hope to see a suggestion of hair on the coat of Mêne's horses and the individual feathers on the wings of Moigniez's birds. The use of both wax and sand in conjunction for casting is known (i.e. sand cast bases and *cire* for the main body of the sculpture.)

Barye used the 'lost wax' method for some work, and it was certainly employed by the founders Gonon, Honoré and his sons. Barye worked closely with this family, and between them they produced much of his important sculpture up to 1839, when he started his own foundry. Some works from this period bear the inscription "Fondu d'un seul jet sans ciselure par Honoré Gonon et ses deux fils". Cast in one pour and without use of chisels. Barye's famous "Lion and Serpent", placed originally in the Tuileries Gardens and his "Tiger and Gavial" which are both now in the *Salle de Barye* at the Louvre, bear this proud inscription, as do some sculptures illustrated in this book (Barye's 'Hunts', Walters Art Gallery, U.S.A.). In the *atelier*, amongst the models in clay and wax, the many drawings pinned up, also the necessary tools, as described by a contemporary, Eugène Guillaume, was "the Master, encircled in his bronze-worker's apron, touching up plasters, chiselling them, putting them in the vice, and examining them from all aspects. His energy was indefatigable, and only when a piece was completely finished would he sign". That Barye's studio was a hive of activity there is no doubt when considering the nature of his work with many small sculptures as opposed to one or two large ones in progress, we are given an insight into the amount of energy needed and work involved. There were various time-consuming skills employed to produce bronze sculpture, whether using wax or sand as a media, the making of the intricate piece-moulds from the originals, the founding, the trimming and the finishing of casts, as well as the allowing for time for the creation of new works. Barye's noble patrons were hardly lavish, and he found it necessary to cultivate a clientele amongst the general public in order to make a living. His workshops were not vast, and even with his pupils he was unable to produce enough without additional help. Barye is known to have employed other craftsmen-founders, also sculptors, to assist in various capacities and also to have supplied models to other foundries to cast which were invariably signed to this effect. It was an economic necessity to make use of them and whilst all work cannot be from 'the hand of Barye' he nevertheless had a finger in the pie — they are 'of the period', and cast in the lifetime of the artist, which is what one looks for in animalier sculpture.

Barye was the 'prototype' whose practice was employed by other sculptors. As well as founding their work, many became their own *éditeurs*. These *éditeurs* were the intermediaries between the artist and the buyer, they handled the contracts for the sculptors, educated the public, and generally encouraged both parties — the publicity agent of today.

Fratin was not a founder but worked closely with the foundries he used, and the quality of his work as a whole is excellent. His early models were cast by Quesnel, who, judging from examples seen, cast work of fine quality, but the use of the *cire* method proved too costly, and Quesnel was forced out of business. Unlike Barye, who set up *ateliers* in several different places and accepted public posts during his career, Mêne 'stayed put', operating his own foundry over a period of forty years, doing the work himself or with the assistance of pupils and craftsmen. Some two hundred groups are known for Mêne, but these include duplicates; either smaller editions or miniatures of the original model, or re-editions of the sculpture (i.e. from a group of animals, removing one or more, and issuing them singly). Mêne presumably made use of the *appareil réducteur,* a mathematically correct contrivance for reducing the size of a sculpture, invented in 1839 by a Monsieur A. Collas. Frémiet operated on a similar basis to Mêne. Secure and settled, he sculpted his small animals for sale to the general public; some are very fine, and some, probably cast by artisans, as they bear no founder's stamp, pay scant attention to detail. This is a pity when one knows that he had good work done by Barbédienne, and also by Eugène Gonon.

Everyone interested in animal sculpture will know that, apart from special commissions or other 'prestige' work, there is usually more than one cast of each group. There are exceptions to this, and it is possible that three of the 'Hunts' in the possession of the Walters Art Gallery are the only ones of their kind. Cast in 'lost wax', they are too complicated for the piece-mould method and were doubtless done on the complete loss of the model basis. They are not listed in any of Barye's catalogues, which were published after the groups were in existence. (As far as I know, no other examples in bronze exist; the Louvre has plasters touched with *cire* of the "Bull" and the "Lion" hunts and their "Indian Mounted on an Elephant" is part only of the "Tiger Hunt"). The question of the numbers cast of each group is difficult to determine, but naturally the more successful and popular models were produced in larger editions than the others. The numbering of casts and the limiting of the editions was not in force until the end of the nineteenth century, when the founder Hébrard purchased the Degas plaster models of horses and ballerinas, and cast in editions up to forty. Hébrard greatly reduced the editions of sculpture later (six or twelve were usual for Bugatti's work), and today the accepted figure is up to eleven only.

After the death of Barye and Mêne, many of their models were bought by Barbédienne and the Susse Frères, who issued catalogues of these works, cast them, and always signed for the foundry. Whilst these were posthumous, they were, with very few exceptions, created from the artist's original model, a very different thing from 'pirate' casts. These are bronzes, cast from worn-out sculptor's models which have had to be retouched, or cast from an actual bronze itself, and are reasonably easy to spot. Altogether poor, unusually heavy, and with much loss of detail, they have dull, lifeless patinas, which show none of the depth and lustre that appears with age. A mould taken from an original bronze sculpture will produce a cast that, as well as being less defined, is smaller than the master.

Some of Mêne's models were made in cast iron with a bronze finish, also some in bronze itself by the Coalbrookdale foundry. Whilst they are not worth anything like the price of a bronze by Mêne many of these replicas are very good indeed, with plenty of detail. Presumably, as they are very exact, the sculptor or his heirs sanctioned the production of these groups and supplied a sculptor's model. The Falkirk foundry, too, made cast iron copies of Mêne's work, but they are inferior to those of Coalbrookdale. As a rule both foundries signed these works. Some cast iron copies of Mêne's work have been made in Russia since the war. (An unknown model of a mare and foal, also "Djinn", are two examples). They are smooth and clumsy, but, as well as having Mêne's signature, they also have some lettering in Russian on the underside of the base. Probably intended as decorative objects only, not fakes, they are easy to spot. Many Spelter or Britannia metal models of animals were made in the nineteenth century. This is the 'poorman's' bronze and is known to most people by now. It can never equal a bronze in finish and detail, being soft metal and usually darkish grey in shade. Some of these models are attractive, but any value that they may have is only decorative.

This section of the book has been based on information available concerning the technical details of founding in the nineteenth century, on experience gained through handling animalier work over some years and on the opinion of a present day expert in casting methods who has personally examined much of the sculpture illustrated. Another source of information is the work of biographers, particularly of Barye, whose techniques are an excellent guide to the general practice employed by the other craftsman-founders.

In animalier sculpture the composition of the group must obviously appeal initially, but the importance of the cast itself cannot be stressed often enough. To recognise the best is not too difficult if careful examination of the sculptures themselves is possible. Good from bad will become apparent, particularly when comparisons are possible.

Author's Note:

The blue/green patination on Mêne's Horse and Jockey illustrated on the back cover is an effect caused by the lighting necessary for colour photography. (The actual patination is very dark black/brown). I must point out that green patinas on the work of both Mêne and Moigniez are invariably suspect.

Barye, however, made much use of green in patinations and seldom used the paler brown shades, seen on some of his earliest work, after the 1840's.

ADDITIONAL PHOTOGRAPHS OF INTEREST

Illustration of a very poor cast of a model originally by Barye (Height about 4 inches). This example lacks depth of patina and detail, and the cast has a coarse, pitted look. The perfunctory patina has disappeared almost completely, and there is little evidence of detailed work in either the mould or the chiselling afterwards. Although the original model was not intended to have the very defined surface detail of a 'realist' work, it should be compared to the two Bulls by Barye illustrated previously.

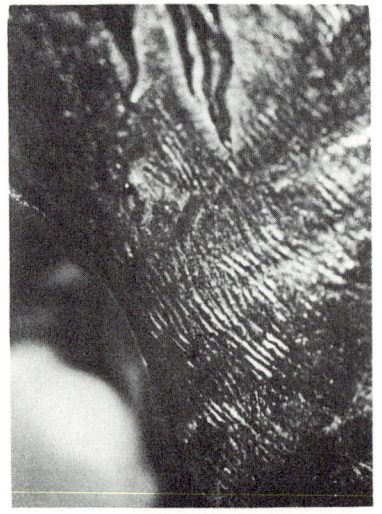

Two examples of the very fine detail achieved on casts by P.J. Mêne, in the mould from the original model and in the chiselling on the bronze itself. (Note the surface of the horse's 'coat'.) In a much used mould, or in a cast taken from a bronze, the work would be altogether less defined, and it would be virtually impossible to add convincingly the meticulous detail of the Jockey's blouse by working on it afterwards. The buttons and buttonholes are particularly successful.

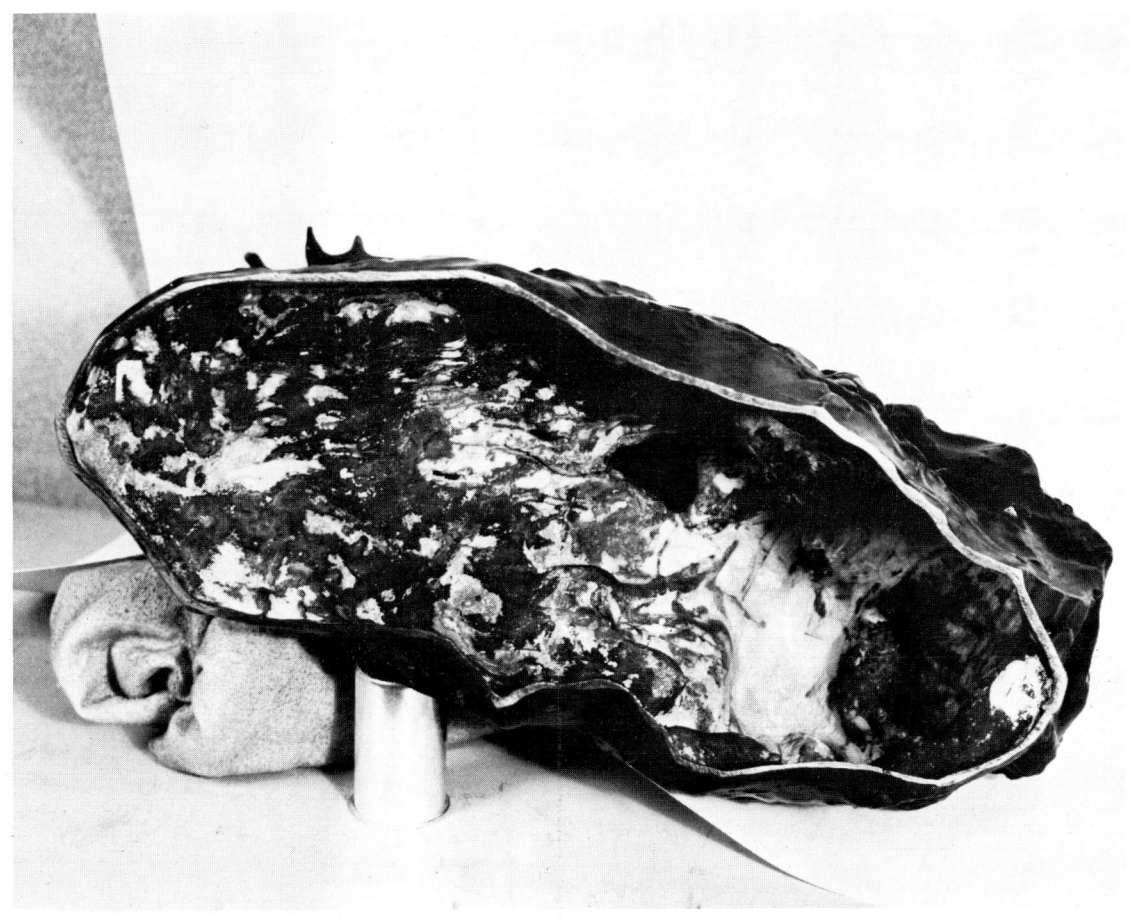

The underside of a cast by 'lost wax' which has not been cleaned, showing the remains of the 'core' adhering.

SIGNATURES OF SCULPTORS

Many artists' signatures appear very clearly on the illustrations included in this book but the following may be of additional help:-

Alexandre Arson — ARSON, or A. ARSON, in capitals, hand-printed, not stamped.

Antoine-Louis Barye — BARYE in capital letters, hand-printed (not stamped individually in Roman letters), is the usual signature. A – L BARYE is another form, but rarely seen, and must not be confused with A. BARYE (the signature of his son Alfred and never used by Barye). A – L BARYE is a smaller signature altogether than the more usual one and appears on certain models only; the "Eagle and Dead Heron" and the "Jaguar and Hare" are two in which the examples examined have always been signed in this way. Sometimes in addition to a normal signature Barye added a stamp BARYE, or with the addition of this stamp, a number on some of his work. There are many theories as to when he ceased this practice, probably by the 1850's, although he mentions it in a letter dated 1859. There is work from Barye, known to be prior to this date, that has neither stamp or number. (123 sculptures were catalogued by the Louvre, of their own and works loaned from other French museums, in the 1956 Exhibition of Barye's sculpture — nine had the addition of the stamp, and one the stamp and number). That Barye ceased this rather erratic practice is certain and all one can be reasonably sure of is that casts inscribed in this way are among those executed in the first half of the century.

Suspect signatures. The A from Alfred Barye's work has been removed leaving simply BARYE — also a reversal of two letters, BAYRE, has been seen.

Alfred Barye — A BARYE, in hand-printed capitals, or A BARYE fils.

Isidore Bonheur — I. BONHEUR, always in small neat hand-printed capital letters. Occasionally Isidore Bonheur in full.

Rosa Bonheur — Rosa B only. Hand-written, script.

Auguste Cain — CAIN or A. CAIN, hand-printed capitals, occasionally the initials A.C.

Paul Comolèra — P. Comolèra, hand-written script.

Paul Edouard Delabrièrre — E. DELABRIERRE, smallish neat capitals.

Alfred Dubucand — A. DUBUCAND, hand-printed capitals.

Christophe Fratin — Fratin signed his surname only, FRATIN, never with the addition of an initial, in small hand-printed capital letters. Sometimes he employed a small stamp (three-quarters of an inch) in lieu of this, again with just the surname (occasionally the N is reversed (И) on this). This stamp is impressed.

Emmanuel Frémiet — E. FREMIET in hand-printed bulky capitals is usual; also a stamp (one inch) with just the surname, impressed, with the letters raised, is sometimes used.

Georges Gardet — G. GARDET, hand-printed capitals.

Joseph Raymond Paul Gayrard — Surname only, hand-printed capitals or script.

Willis Good — J. Willis Good, or Good, hand-written script.

Henri-Alfred-Marie Jacquemart — A. Jacquemart, hand-written script or the initials A – J in capitals.

Pierre Lenordez — P. Lenordez, hand-written script.

Pierre Jules Mêne — P.J. MÊNE in bold Roman capitals (possibly stamped individually) is the usual signature for this artist. (Never simply the surname or initials alone). He also used a smaller capital letter signature P.J. MÊNE. It is neater and more uniform than his normal rather cumbersome one. Rarely seen, it is on small works only.

Jules Moigniez — J. Moigniez, hand-written script.

Le Comte du Passage — Ct. du Passage, hand-written script, not capitals.

Ferdinand Pautrot — PAUTROT or F. PAUTROT, capitals.

A very rare signature for Rosa Bonheur, (Rosa B is usual). Detailed photograph from her large dated sculpture, illustrated earlier.

KNOWN FOUNDERS OF ANIMALIER SCULPTURE

Ferdinand Barbédienne

Barbédienne was one of the most important nineteenth century founders. Opening his factory in 1838, he cast many bronzes, those of contemporary, as well as of earlier artists. His splendid business acumen and great personal taste did much to propagate art and its appreciation. He cast some of Barye's models in the early 1850's and after the artist's death bought a great many of his models, re-editing them.

Casts from the Barbédienne foundry were signed, "F. Barbédienne." It is not certain at what period those bearing the initials "F.B.", or the circular stamp "Collection Barbédienne" in gilt, were made. That Barbédienne founded a collection of Barye's work is known, and these particular casts may be connected with this in some way. (Signature illustrated).

Boyer and Gayrard

Both listed as nineteenth century sculptors. Known to have worked together as sculptor-founders and to have been involved in work for Barye.

Coalbrookdale Company, England

This foundry dates back to the seventeenth century. They built the first cast-iron bridge in 1779 and in the nineteenth century they achieved a reputation casting works of art in both iron and bronze. Some of these works are of exceptional quality and finish and it is known that many were made as exhibition pieces for the Great Exhibition of 1862 to prove the worth of English casting. The models in bronze bear the signature "Coalbrookdale — bronze" and sometimes the full signature of the original artist.

Alfred Daubré

Established as a founder in 1855, he edited some of Fratin's later work.

Durand (Eck et Durand)

Eck not known; Durand possibly Amédée-Pierre, who had a successful career as a sculptor (he won the Prix de Rome as a medallist). He gave up sculpture for industrial and agricultural interests, but later inherited a well-known *atelier* from an uncle.

Debraux d'Englure

The term *éditeur* was first used in connection with this founder in 1839.

Fumière

Active in the 1850's.

Honoré Gonon
One of the foremost founders of the first half of the nineteenth century. A contemporary of Barye, he cast much of his important work using the *cire perdue* method.

Eugène Gonon
Son of Honoré, born in 1814, sculptor and founder. Pupil of Prodier and Blondell. He employed the *cire perdue* method of casting used by his father. Known to have cast work of Barye and Frémiet.

Hippolyte Peyrol
Married Juliet Bonheur (1830–1891) sister of Rosa and Isidore, in 1852. Exhibited work at the Great Exhibition of 1862 in London. Edited the sculpture of Rosa and Isidore Bonheur (signature illustrated).

E. Quesnel
This was a small firm of founders in Paris working between 1811 and 1847. Some of Fratin's models were cast at this foundry; edited by the artist, they were cast at his expense.

E. Siot-Decauville.
Member of Society of French Artists. Became *Chevalier* of the Legion of Honour. Later in his life he concerned himself with editing works of sculpture. Died 1909. (Much of Gardet's work bears his stamp).

Susse Frères
Susse Frères are amongst the most prominent of the nineteenth century founders and editors. J.V. Susse (1806–1860) and J.B.A. Susse (1808–1880), did much to popularize the small bronze. After a period of editing in plaster, they held the first exhibition of their work in 1839, and in 1840 they organized a large bronze workshop, fifty per cent of this work being exported. Many bronzes cast by Susse have a rich mid-brown patination which is very distinctive.

Henri Léon Thiébault (or Thiébaut)
1855–1899. Sculptor and founder.

Pierre Phillipe Thomire
1751–1843. Started his own workshop in 1776. Had a great reputation as a founder to the court. Retired in 1823, entrusting the firm to his associates, when it became Thomire et Cie. (later Thomire-Dutherme et Cie.).

APPENDIX A – METHODS OF CASTING

Among the various methods of casting bronze sculpture recorded throughout history, there are two that are predominant. The *cire perdue* (lost wax) process, with its variants, which resulted in the loss of the artist's original model until the piece-mould was evolved; and the method of casting with the use of sand. This again involves the piece-mould, but, however this is used, its important function is that it can be taken apart and re-used (to a limited degree).

LOST WAX

The first use of the *cire perdue* process, recorded as much as a thousand years or more B.C., resulted not only in the loss of the original model, but in one cast only. The sculptor modelled in wax, which was then completely encased in a heat-resisting material known as the 'waste' mould, which had necessary vents (runners and risers) to allow the wax to be melted out and the molten bronze to be poured in. When this was done and the bronze had cooled, the mould was broken away and a solid bronze sculpture was the result. The next step was similar, except that the original wax model contained a refractory material 'core', pinned in position through the wax to the mould, which remained after melting out the wax and the bronze 'pour'. This could be scraped out after cooling, leaving a hollow, not solid, bronze cast. This cast would be lighter and altogether more practical for large works, but it still only allowed for one cast from one original model.

More than one cast of a work could be made when a piece-mould was devised. This method originated in Greece and was known to be in use by the end of the fifteenth century in Italy. This piece-mould was made by encasing the original model in a light plaster cast which was carefully cut into sections to permit taking apart and fitting together again. In the fragile mould so formed, several or many copies of the original could be cast in wax, (with or without a core, according to the design of the model). The wax model had then to be cleaned up where the joints in the mould occurred and often at this stage the finer details were added by the artist. The wax model was then enclosed in a block of ceramic material, (the waste mould) and subsequently melted out. The bronze was then poured into the mould, provision having been made for runners which allow the metal to be poured in, and risers which allow the air to escape and the metal to flow to all points. The mould then had to be broken off the bronze; thus only one cast was made from a wax and its mould; because of this there can usually be found small differences between them and each bronze therefore becomes an individual work. The bronze cast broken out of the waste mould had still to be worked on. The necessary supports and the runners and risers had all to be chiselled off and any other flaws or chemical deposit dealt with. Often further chiselling of detail was done at this stage prior to patination and when all these processes are carried out by the artist himself, the bronze is 'autograph' in the true sense.

SAND

Very simply, sand casting involves the use of boxes (or flasks), which contain the sand which forms the sections of the moulds of the bronze to be cast. Sand, if it is the fine sea-shore variety, which gives sharper detail to these moulds, has to have the addition of various bonding aggregates — fuller's earth, and linseed, vegetable or fish oils are all good bonding agents. Quarried sand which is naturally bonded with clay can also be used, but would make a rougher mould, needing more finishing.

The construction of the boxes in which sand casting is carried out makes the production of a single piece shell an impossibility. The pressures when tamping to consolidate the moulds — also the 'withdrawal' difficulties — preclude this, and a variation of the piece-mould is used. This piece mould differs considerably from the one employed in the *cire perdue* method, as it is not the 'negative' impression, (the moulded sand is this), but the master, or positive form. This master is an exact replica of the sculptor's original model, in a substance such as clay or terra-cotta that can withstand pressure, but made in several divisible parts that have been carefully cut and trimmed to fit properly one into another. It is hollow except for the thinner extremities, legs, heads, tails, etc. The size and shape of these sections determine the eventual thickness and the individual pieces of the bronze cast, as, when they have been used to make the sand impressions (negatives) on both their outer and inner surface, the sections are removed (to be used again); then the sand moulds are carefully pinned back into place with a gap where the master has been and where the bronze 'pour' can enter. When cast, these sections can be assembled to form the finished bronze. Much skill is needed in the original division of these master models; there is usually a large part that is cut to include the main body of the sculpture, with smaller segments undercutting it. Thus, in the casting of a simple animal form, it is possible to include in the main piece the whole of the upper part and sides of the body, together with the head, neck, shoulders and legs. In other words, as much can be included as can be pressed into the sand and then withdrawn without destroying the impression made. Additional sections that have been under-cut are pressed separately to form the other sand moulds necessary to complete the work, and additions such as reins, spears and other appendages of this sort are not cast, but added afterwards. Although sand casting was a quicker method than the *cire perdue*, as the contents of a number of boxes could be 'poured', enabling several casts to be progressing at the same time, the assembly of the separate pieces was intricate and exacting, as was the chiselling necessary to 'finish' the work.

APPENDIX B – LISTS OF SCULPTORS' WORKS, SALON EXHIBITS ETC.

ANTOINE-LOUIS BARYE
LIST OF ANIMAL SCULPTURE

This comprehensive list of Barye's work was compiled by Arséne Alexandre (Animal and Equestrian sculpture only has been listed here).

Tiger and Gavial (Louvre)
Lion and Serpent (Jardin des Tuileries – Louvre)
Lion – Bas-relief, July Column
Lion Resting (Carrousel)
Tiger Devouring a Stag (Museum of Marseilles)
Theseus Fighting the Minotaur
Centaur and Lapith
Theseus Fighting the Centaur Biénor
Centrepiece (for the Duke of Orleans)

The Duke of Orleans
General Bonaparte
Gaston de Foix
Charles VIth in the Forest of Mans.
Charles VIIth
Tartar Warrior Reining his Horse
Two Arab Hunters Spearing a Lion
Hunter Surprised by a Serpent

Elephant Ridden by an Indian (Louvre)
Caucasian Warrior
Huntsman in the Costume of Louis XVth
Angélique et Roger
Orang-Utang Riding a Gnu

Bear Overcome by Hounds
Bear Fleeing from Hounds
Two Young Bears Fighting
Bear Eating an Owl
Standing Bear
Seated Bear
Ratel Discovering Eggs

Greyhound Resting
Tom, Large Greyhound of Algeria
Greyhound and Hare
Pointer ("Braque") and Pheasant
Setter and Pointer and Partridge
Setter ("Epagneul") and Pheasant
Seated Basset (Dachshund)
Standing Basset (Dachshund)
English Basset (Dachshund)

Wolf Attacking a Stag by the Throat
Wolf Abandoning its Prey
Wolf Caught in a Trap

Two Young Lions Fighting
Lion Holding an Antelope
Lion Devouring an Antelope
Lion and Serpent (Study for the Tuileries).
Seated Lion
African Lion
Algerian Lioness
Walking Lion
Walking Tiger
Tiger Surprising an Antelope
Panther Striking a Stag
Tiger Surprising a Stag
Tiger Devouring a Gazelle
Panther Resting
Panther of India
Panther of Tunis
Panther Surprising a "Zibeth" *(Civet Cat)
Panther Holding Down a Stag

---

* This is listed as "Zèbre", probably a printer's error. There is no known work by Barye, or catalogued by him, which includes a zebra – Author.

Jaguar Devouring a Hare
Walking Jaguar
Standing Jaguar
Jaguar Holding Down an Alligator
Jaguar Devouring an Agouti (Rodent)
Sleeping Jaguar
Jaguar Devouring a Crocodile
Ocelot Carrying away a Heron

Cat
Rabbit
Seated Hare
Frightened Hare (Rabbit, Ears Lowered?)

Elephant Crushing a Tiger
Elephant of Indo-China
African Elephant (Senegal)
Indian Elephant (d'Asie)
African Elephant

Horse Surprised by a Lion
Thoroughbred Horse
Part-bred Horse
Part-bred Horse Head Lowered
Cheval Turc
Cheval Percheron
"Hémione"

Algerian Dromedary
Harnessed Egyptian Dromedary
Dromedary Ridden by an Arab
Persian Camel

Elk Surprised by a Lynx
Family of Deer
Ten-pointed Stag Brought Down by a
    Scotch Hound
French Stag Walking
French Stag in Repose
Listening Stag

Braying Stag
Stag with Leg Raised
Family of Stags
Stag Rubbing its Antlers on a Tree
Axis
Java Stag
Axis Stag
Stag of the Ganges
Virginian Stag
Dead Bouquetin
Ethiopian Gazelle
Kevel

Bull
Rearing Bull Struck at by a Tiger
Bull Brought Down by a Bear
Little Bull
Buffalo
Wounded Boar

Eagle Holding a Heron
Eagle with Wings Outstretched
Eagle Holding a Snake
Parrot in a Tree
Common Pheasant
Wounded Pheasant
Golden Pheasant
Stork on a Turtle
Owl
Marabout Stork
Turtle
Crocodile
Crocodile Devouring an Antelope
Python Devouring a Doe
Python Crushing a Crocodile

Buck
Doe and Fawn
Resting Doe
Resting Hind
Fawn
Group of Rabbits

Elk Surprised by a Lynx
Python Striking a Gnu at the Throat
Tiger Devouring an Antelope
Horse Attacked by a Tiger
Deer Brought Down by Three Algerian Hounds
Deer Overthrown by Two Hounds
Lion Devouring a Boar
Seated Bear
Pheasant in a Tree
Dead Gazelle

Bear on its Back
Crouching Panther Holding a Gazelle
Head of a Chimpanzee

Bas-reliefs:-
Lion of the July Column (Zodiac?)
Leopard
Panther
Genette Carrying Away a Bird
Virginia Stag

This list has been checked with the five catalogues issued by Barye, which contain additional items, either decorative works, or variations of the same model on a different scale, reversed to form a pair, or re-edited. (i.e. the "Rearing Bull" is listed as "Taureau Cabré" in preparation (without the Tiger) in Barye's first catalogue. There are two Rabbits, Nos. 3 and 4 — Ears Lowered and Raised in his first catalogue and five Bassets, two Standing (pair), and two Seated (pair), and "Petit Chien Basset Debout" in his fourth catalogue). There are two groups which do not check precisely with Barye's catalogues: i.e. There is only one "Elk attacked by a Lynx" listed, and no record of a Deer brought down by *three* Hounds.

CHRISTOPHE FRATIN'S EXHIBITS AT THE FRENCH SALON

1831	Fermer, English Thoroughbred
1831	Two Bulldogs Playing with a Greyhound – wax
1833	Greyhound – study in plaster
1833	Chained Bulldog – plaster
1833	*Ecorché* Horse – bronze
1834	Wild Horse Attacked by a Tiger
1834	Canadian Stag Cornered by Dogs
1834	Felix, Thoroughbred from the Viroflay Study
1834	Panther Seizing a Gazelle – plaster
1834	Dead Cow being Devoured by Wolves – plaster
1835	A Greyhound: A Bulldog – bronze commissioned to ornament the entrance to the factory of M. Brame-Chevalier at Montesson near St. Germain-en-Laye. These bronzes were cast in 1835 at the *atelier* of Quesnel, 13 Rue des Enfants-Rouge, Paris.
1835	Vulture Devouring a Gazelle – bronze
1835	Tiger Bringing Down a Young Camel – bronze
1835	Lion Devouring a Zebra – bronze
1835	Elephant Killing a Tiger – bronze
1835	Lioness Bringing her Prey to her Cubs – bronze. Terra-cotta example Musée de Compiègne
1835	Reclining Stag Licking Itself – bronze
1835	Dead Horse – bronze. Cast by Quesnel
1835	Rinbow, Stallion – bronze. The property of M. Reinssec. Cast by Quesnel
1835	Bull Fighting with Dogs – plaster
1835	Three Racehorses – plaster
1836	Two Dogs – bronze. These bronzes were donated by the artist for the Esplanade at Metz in 1836
1836	Lion Carrying Away its Prey – bronze
1836	Bulldog and Greyhound. These groups were bought in 1836 by an English M.P., Lord Powerscourt.
1836	Tiger Seizing its Prey – plaster.
1837	Lion (refused Salon)
1837	Horse at a Gallop (refused Salon)
1837	More and Foal – plaster
1839	Eagle and Vulture Disputing Prey – bronze.
1839	Two Stags in Battle (refused Salon)
1839	Lion Bringing Down a Stag (refused Salon)
1840	Lions to ornament the pediment of the Château de Dampierre belonging to the Duk de Luynes, also brackets supported by Panthers for the same Château.
1842	Tony – bronze horse. Musée de Compiègne
1850	Thoroughbred Horse – bronze. Commissioned by the Ministry of the Interior in 1848. Later placed on the Esplanade at Metz.
1850	"Study" – terra-cotta.

1850's	Oval plaquette containing four heads of animals, Monkey, Cat, Dog and Tiger. Musée de Tournus 'Saône et Loire). (Donated in 1879).
1850	Two works depicting Various Heads
1852	Triumphant Eagle – plaster
1852	Eagles – bronze group for Esplanade at Metz commissioned by the Ministry of the Interior. (Plaster model, Salon of 1852)
1853	Horse Attacked by a Lion – bronze. Commissioned by the Ministry of the Interior for the Square Montrouge, Paris. (Plaster model, Salon of 1853)
1858	Tiger – terra-cotta. Acquired by the Société des Amis des Arts de Bordeaux in 1858.
1860	Combat, Lion and Tiger – plaster. Commissioned 1860. Musée de Metz.
1861	Recumbent Stag, Calling – possibly the same as Stag at Bay.
1861	Stag at Bay – bronze. Esplanade at Metz.
1862	Nanny-goat and Horse – plaster
1862	Arab Horse – plaster. Musée de Châlons-sur-Marne.

P.J. MENE'S EXHIBITS AT THE FRENCH SALON, OTHER EXHIBITIONS, AND IN MUSEUMS

1838	Salon	Dog and Fox – bronze
1840	"	Horse Attacked by Wolf – bronze
1840	"	Several Animals – bas-relief in terra-cotta
1840	"	Stag – bronze
		(Possibly the same as Stag Browsing (bronze) in the Musée de la Rochelle)
1840	Salon	Merino Ram – bronze
1841	"	Fox "d'Islande" and Cock – bronze
1841	"	Panther of Constantine and Gazelle – bronze
1842	"	Panther Throwing itself upon a Roebuck – bas-relief terra-cotta
1842	"	Racoon Seizing a Duck – bronze
1842	"	Stag "Cerf Muntjac" – bronze
1842	"	Roebuck – bronze
1842	"	Fox "Renard d'Islande" – bronze
1843	"	Jaguar and Alligator – bronze
1843	"	Ibrahim "Cheval Arabe Ramené d'Egypte" – bronze
1843	"	Pointer Carrying a Hare – bronze
1843	"	Cross-bred Spaniel – bronze
1843	"	English Basset (Dachshund)
1844	"	Stag Hunt – bronze
1844	"	Brazilian Jaguar – bronze
1844	"	Hind – bronze
1845	"	Normandy Bull – bronze
1845	"	Cow and Calf – bronze
1845	"	Ewe and Lamb – bronze
1845	"	Ram – bronze
1845	"	Goat – bronze
1845	"	Spanish Greyhound "Large Species" – bronze
1845	"	Greyhound and Hare – bronze. (Musée de la Rochelle)
1846	"	Pure-bred English Spaniel – bronze. (Musée de Marseilles and Musée de la Rochelle)
1846	"	Hare (Musée de la Rochelle)
1848	"	"Chasse à la Perdrix" – wax
1850	"	" " " – bronze
1848	"	Vixen and Young – wax
1848	"	Half-bred Ram – bronze
1848	"	Two Greyhounds – plaster
1849	"	Fox Hunt – wax
1849	"	Study of Game – wax
1849	"	"Djinn", Barbary Stallion – bronze
1850	"	Kemkon-Handani, Arab Mare and Foal – wax
1850	"	Pedigree English Pointer Guarding Game – wax

1850	"	"Chasse au Canard", Two Retrievers and Duck – bronze
1850	"	Stag and Hind – wax
1850	"	Running Thoroughbred Horse – plaster (Musée de Marseilles)
1852	"	Tachiani et Nedjibé, Arab Horses – wax
1855	Exposition Universelle	" " " " " "
1852	Salon	Dead Roebuck and Heron – wax
1852	"	Dog and Game – bronze
1853	"	Battling Stags – wax. (Donated to the Musée Louvre by Mme. Cain)
1853	"	Arab Horses – bronze
1855	Exposition Universelle	"The Kill", on Foot – wax
1855	" "	Terriers – wax
1855	Salon	" "
1857	"	Stag Hunt – bronze
1857	"	English Dogs or Hounds – bronze
1857	"	Basset Hounds Hunting in Undergrowth – wax
1857	"	Dog "Race Française" – wax
1859	"	" " " – bronze
1859	"	Roe Deer – wax
1859	"	Mare and Dog – bronze
1859	"	Bitch and Puppies – plaster
1861	"	"La Prise du Renard, Chasse en Ecosse" – wax
1862	Exposition de Londres	" " " " " " " – silver
1867	Exposition Universelle	" " " " " " " – "
1861	Salon	Roebuck – bronze
1861	"	Dead Hare and Fish – still life
1862	Made for the collection of the Duc d'Aumale	Boar Hunt – "the Kill"
1863	Salon	"Vainqueur du Derby" – wax
1864	"	" " " – bronze
1863	"	Fox Hunt – "The Kill" – silvered bronze
1863	"	Breton Horse – bronze
1864	"	Huntsman and Bloodhound – wax
1865	"	Amazone – English Thoroughbred Horse – wax
1867	"	" " " " – bronze
1865	"	Mountain Sheep – wax
1866	"	"Winner of the Race" – wax
1866	"	Fox and Pheasant – wax
1867	"	Pointer
1868	"	Mare and Foal – wax
1869	"	" " " – bronze
1878	Exposition Universelle	" " " "
1869	Salon	Mounted Huntsman Leading his Hounds – wax
1870	"	" " " " " – bronze
1872	"	Huntsman of the Period of Louis XVth – bronze
1878	Exposition Universelle	" " " " " " "

1872	Salon	Beagling "Un Chasse au Lièvre" – bronze
1872	"	Three Terriers, "Chasse au Lapin" – bronze
1900	Centenary Exposition of French Art.	" " " " " " The wax model given to the Louvre Museum by Mme. Cain. Bronze – Musée de Marseilles)
1873	Salon	Falconing "Chasse au Faucon" – wax
1874	"	" " " " – bronze
1878	Exposition Universelle	" " " " "
1876	Salon	Picador – wax
1877	"	" – bronze
1878	Exposition Universelle	" "
1877	Salon	Toreador – wax
1878	"	" – bronze
1878	Exposition Universelle	" "
1878	Salon	African Hunter – wax
1879	"	" " – bronze
1879	"	Valet de Limiers – wax

LIST OF P.J. MENE'S KNOWN WORKS

Figures and Animals
Huntsman of Louis XVth and his Pack.
Hunt in Scotland (La Prise de Renard).
Picador on Horseback
Arab Falconer on Horseback
Huntsman of Louis XVth on Horseback
Huntsman of Louis XVth on Horseback with Hounds at Bay
Huntsman and Bloodhound
African Huntsman on Horseback
Toreador – Matador on Foot
Scottish Huntsman with Two Hounds
Arab Falconer on Foot
Scotsman with Hound Holding up a Fox
Jockey Holding Derby Winner
Horse and Jockey

Hunts
Stag Attacked by Four Hounds
Stag Attacked by Three Hounds
Boar Attacked by Four Hounds
Fox Held by Two Hounds
Two Setters Hunting a Duck
Hare Hunted in Undergrowth
Three Terriers around a Rabbit Hole
Pointer and Setter 'Setting' a Partridge

Horses
Mare (Hunter) with a Highland Terrier
Arab Mare and Stallion
Normandy Mare and Foal
Normandy Mare
Arab Mare and Foal
Arab Mare (Nedjibé) Saddle and Gun at her Feet
Riding School Mare Saddled and Bridled with a Dog
Racehorse – Saddled and Bridled
Horse Attacked by a Wolf
Horse at Liberty
Arab Mare Saddled (Nedjibé)
Arab Horse Tethered
Horse at the Fence (Djinn)
Breton Horse
English Mare (Rédinha)
Arab Horse (Ibrahim)
Percheron
Frightened Horse

Dogs
Two Pointers Resting
Pointer Guarding Game
Bassets Hunting in Undergrowth
Bitch and Puppies
Pointer and Setter in Undergrowth
Two Small Dogs at Bay
Setter Retrieving Duck
Pointer Retrieving a Hare
Griffon with Two Pigeons
Retriever with a Whip (Warwick)
Bloodhound
Pointer Carrying a Hare
Setter (Sylphe)
French Dog (Belotte)
Pointer (Trim)
Seated Pointer Guarding a Rabbit
English Setter (Médor)
Dog Attacking a Fox
Basset or Dachshund (Legs Twisted)
Panting Hound (Milla)
Pointer (Marly)
Pointer (Tac)
Pointer (Tom)
Pointer (Low)
Pointer (Without Hare in its Mouth)
Pointer with Duck
Retriever – French Spaniel (Fabio)
Greyhound (Spanish)
Spanish Greyhound with Hare
Greyhound and King Charles Spaniel
Dog (Mignonette)
Two Greyhounds Playing with a Ball
Greyhound with a Ball (Jiji)
Smaller Greyhound with a Ball (Gizelle)
Seated Griffon
Long-haired Terrier Lying Down
Griffon (Casca)
Basset (Trompette)

Dogs (cont'd).

Spaniel or Setter (Sultan)
Hound (Wagram)
Wounded Hound
Terrier Ratting (Tom)
Terrier Licking its Paw
Greyhound and Fan
Spaniel (Diane)
Dog (Lutine)
Dog (Frisette)
King Charles Spaniel

Deer

Two Stags Fighting
Stag Browsing at Branch
Stag, Stabbed
Stag Rubbing against a Tree
Common Stag
Muntjac Stag
Frightened Stag
Stag Head Turned
Group of Stags
Group of Deer
Deer Going Forward
Deer Lying Down
Frightened Hind
Frightened Hind Lying Down
Axis (Roe deer) Lying Down

Farm Animals

Cow and Calf
Normandy Bull
Half-bred Ram
Sheep Feeding its Lamb
Mountain Sheep Standing
Mountain Sheep Lying Down
Billy-Goat
Nanny-Goat and Kid
Family of Goats
Browsing Goat
Indian Goat
Goat Lying Down
Goat, Leg Raised
Baby Goat

Various Additional Subjects

Two Urns, Decorated with Hunting Subjects
Jaguar and Alligator
Group of Roe Deer
Roe Deer on the Alert
Roe Deer Drinking
Roe Deer
Young Roe Deer (Carabit)
Foxes and Pheasant
Family of Foxes
Two Foxes
Arctic Fox and Cock
Arctic Fox Seated
Panting Fox
Standing Fox
Fox at a Fence
Fox Lying Down
Fox
Boar
Leaping Chamois
Racoon Killing a Duck
Cat and Kittens
Cat
Hen and Chickens
Tufted Fowl
Hen
Pair of Rabbits
Rabbit in Repose
Rabbit Outstretched
Pair of Ducks
Duck and Ducklings
Duck Going Forward
Duck Drinking
Dead Duck
Group of Pigeons
Panther and Gazelle
Panther
Jaguar of Brazil
Algerian Gazelle
Desert Gazelle Drinking
Pheasant Going Forward
Pheasant Turning
Cock Standing
Hare in Repose
Partridge
Dead Partridge

Various cont'd.

Ram Lying Down
Nest of Birds
Nest of Squirrels

Plaques. "Still Life" Compositions
Roebuck and Heron
Stag's Head
Fox
Hare and Pheasant
Hare and Fish
Duck and Woodcock
Dead Pheasant
Dead Hare
Dead Cock
Dead Rabbit

AUGUSTE CAIN'S EXHIBITS AT THE FRENCH SALON

1851	Egyptian Vulture
1851	Group of Frogs
1852	Ibis
1853	Family of Partridges
1853	Snipe
1863	Falcon Hunting Rabbits – Bas-relief
1863	Fox Hunting Ducks – Bas-relief
1864	Sahara Lion – plaster
1864	Cockfight
1865	Vulture – Musée de Luxembourg
1867	Family of Tigers
1868	Lions – Tuileries entrance
1874	Pheasants and Nest
1876	Tigers Fighting – Tuileries Gardens
1882	Lion and Lioness Fighting over a Wild Boar – Owned by the State
1883	French Cock
1884	Rhinoceros Attacked by Tigers – Tuileries Fountain.

AUGUSTE CAIN
LIST OF WORK (EXCLUDING MONUMENTAL SCULPTURE)

Cock Fight
Lion and Lioness Fighting over a Boar
Tigress Carrying a Peacock to her Cubs
Pheasants – Plaque
Fox Seizing a Duck – Plaque
Eagle Seizing a Partridge – Plaque
Falcon Flying at a Rabbit – Plaque
Nest of Pheasants
Large Pheasant
Two Pheasants
Group of Pheasants
Cock Pheasant
Little Pheasant
Relay of Hounds (Couple)
Group of Hounds (English)
Group of Hounds (St. Hubert)
Hound (Caron)
Trocadero Bull
French Cock Crowing
Family of Partridges
Partridge
Ibis Hunting a Frog
Cock and Hen
Group of Cock with Hens
Indo-Chinese Cock
Dead Woodcock
Dead Partridges – Plaque
Dead Duck – Plaque
Sparrow Caught in a Trap
Donkey with Panniers
Donkey
Stork and Turtle
Two Ducks
Family of Rabbits
Startled Rabbit
Rabbit and Carrot
Rabbit and Dog
Hen Pheasant
Hen and Chicks
Indo-Chinese Hen
Tufted Hen
Common Hen
Cock on a Basket
Lioness Carrying a Young Wild Boar to her Cubs
Lioness Holding a Lion Cub in her Mouth
Lion and Ostrich
Various Candelabra and ornamental objects decorated with a variety of birds and animals.

SALON EXHIBITS AND OTHER WORKS BY EMMANUEL FREMIET

1843	Salon	Gazelle – plaster
1846	"	Study of a Dog – Terra-cotta
1847	"	Dromedary – wax
1848	"	Carriage Hounds – plaster
1848	"	A Cat – plaster
1848	"	A Fox – plaster
1849	"	Deer Hunter – plaster
1849	"	Family of Cats – plaster
1855	Exposition Universelle	" " " – marble. (Commissioned by the Minister of the Interior).
1849	Salon	Tartar Camel – plaster
1849	"	Heron – bronze
1850	"	Wounded Dog – bronze
1855	Exposition Universelle	" " " (Commissioned by the Minister of the Interior. Founded by Eugéne Gonon).
1850	Salon	Female Cat – marble. (Commissioned by the Minister of the Interior).
1850	Salon	Wounded Bear – plaster
1850	"	Griffon Bitch – plaster
1850	"	Griffon – plaster
1855	Exposition Universelle	" – bronze
1850	Salon	Cat Study – plaster
1850	"	Hen from Indo-China – plaster
1850	"	Marabou Holding a Cayman in its Feet – bronze. (Commissioned by the Minister of the Interior 18th May, 1849).
1853	"	The Killing of a Draught-horse – plaster. (Commissioned for the Veterinary School at Toulouse).
1853	"	Ravageot & Ravageole, Basset Hounds – bronze. In the Salle des Gardes, Palais de Compiègne.
1855	Exposition Universelle	Artilleryman on Horse – bronze. Museum of Rennes
1855	" "	Policeman on Horse – plaster
1855	" "	Barge Horses – bronze. (Museum of Mans)
1855	" "	Dromedary – plaster
1855	" "	Camel – plaster
1855/60		Lions. For the roof of the entrance of the Carrousel.
1857/58	Louvre Museum	Three capitals dedicated to falconry, stags and horses
1859	Salon	Showman's Horse – bronze, plaster
1859	"	Forester, Hunter from the stable of M. le Comte d'Oultremont – bronze, plaster
1859	"	Horse with a Crow – bronze
1859	"	Tethered Horse – bronze
1859	"	Arab Horse – bronze
1861	"	The Centaur Teree Carrying to his Lair a Bear taken in the Mountains of the Hemus – plaster. (Property of Prince Napoleon).

1863	Salon	The Centaur Teree Carrying to his Lair a Bear taken in the Mountains of the Hemus — bronze. (Bonnat Museum, Bayonne).
1861	"	Two Month Old Cat — silvered bronze
1867	Exposition Universelle	" " " " " "
1863	Salon	Gallic Horseman — plaster
1864	"	" " — bronze
1867	Exposition Universelle	" " " (Museum of St. Germain-en-Laye)
1866	Salon	Roman Cavalier — bronze
1867	Exposition Universelle	" " "
1865		Arab Chief on Horseback — bronze. (Collection Frémiet - Barbédienne, Museum of Abbeville).
1864	Salon	Pan and a Bear — plaster
1867	"	" " " " — marble. (Musée de Luxembourg)
1868	"	Napoleon I on Horseback — plaster. (Museum of Grenoble)
1868	"	Neptune Reborn, on Horseback — plaster
1869		A Bull, an Eagle, a Marabou, a Griffon
1869	"	Ox/Bull and Griffon — sandstone
1870	"	Louis d'Orléans, Brother of Charles VI, on Horseback — bronze. (Smaller model in bronze at Museum of Orleans)
1870	Five groups for a fountain designed by Lefuel in Pomerania	Women and Animals
1870	Salon	Seahorses, Dolphins and Turtles. (Fountain of the Observatoire, Luxembourg Square, Paris).
1873	"	Falconer — silvered bronze
1874		Joan of Arc Equestrian — gilded bronze. (Place du Rivoli, Paris)
1876	"	Retiary and Gorilla — Terra-cotta
1876		Jaguar and Gorilla. (Collection of Frémiet - Barbédienne
1878	"	Errant Chevalier — plaster. (Reduction in bronze in Collection Frémiet — Barbédienne
1878		Elephant — bronzed cast iron. (Trocadero Gardens, Paris)
1880	"	Capture of a Young Elephant — marble
1889	Exposition Universelle	" " " " " — plaster
1900	Centenary Exhibition of French Art	" " " " " — terra-cotta
1881	Salon	The Great Condé, Equestrian — bronze (Chantilly)
1882	"	Stefan-al-Mare, Prince of Moldavia, Equestrian — bronze
1883	"	Lamplighter on Horseback (XVth century) — plaster. (Hotel de Ville, and Petit Palais, Paris)
1883	Exposition Universelle	Lamplighter on Horseback — plaster
1889	" "	" " " "
1900	Centenary Exhibition of French Art	" " " "
1883		Marabou and Snake — bronze. (Hotel de Cassin, Paris)
1884		Bear, Lion — marble.
1885		Snake — wood. (Hotel Gérôme, Paris)

1885		Monkey and Butterfly — bronze. (Hotel Dieulafoy, Bayonne)
1885	Salon	Racehorses — bronze
1889	Exposition Universelle	" "
1885	Salon	Bear and Stone Age Man — plaster
1889	Exposition Universelle	" " " " " (bronze — Natural History Museum
1886	Salon	Running Dogs — bronze
1886	"	Greyhounds — bronze
1887	"	Gorilla — plaster
1889	Exposition Universelle	" "
1900	Centenary Exhibition of French Art	" " (Museum of Nantes)
1889		Cat with Chicken — plaster
1900	Centenary Exhibition of French Art	" " " "
1889		Basset Hounds and Cat
1889	Exposition Universelle	Head of a Bull.

EMMANUEL FREMIET

Catalogue of work issued from No. 42 Boulevard du Temple, Paris, 1859–1860. * (Later he had his *atelier* in the Faubourg Saint-Honoré, but issued no catalogue from there, as far as is known.)

Seated Dog
Heron Poised
Farmyard Fowl
Group of Fowl and Rat
Little Cat Playing
Sheep
Awakening Terrier
Ass
Hound Lying Down
Fox Poised
Goat and Kids
Pony
Family of Cats (large)
Family of Cats (small)
Kid
Ducks
Curlew and Frog
Camel
Dromedary
Cock Crowing
Bulldog
Head Study – Basset (large)
Standing Heron
Brood of Indo-Chinese Pheasants

Head Study – Cat (large)
Head Study – Goat
Head Study – Terrier
Fawn
Cat and New-born Kittens
Vagabond Cat (large)
Vagabond Cat (small)
Barge Horses
Griffon and Puppies
Group of Basset Hounds (large)
Jaguar and Monkey
Stag
Gazelle and Jaguar
Cat on Foot
Fox of Egypt
Diving Duck
Bearded Griffon
"Off to the Hunt" (dog)
Group of Basset Hounds (small)
Hunter at the Trot
Hunter at the Gallop
Horse in Hand, and his Groom
Tethered Horse
Part-bred English Horse
Arab Horse

*Animal sculpture only is included here.

ISIDORE BONHEUR'S EXHIBITS AT THE FRENCH SALON AND OTHER EXHIBITIONS

1848	Salon	African Horseman Attacked by a Lion – plaster
1850	"	Bulls – plaster
1850	"	Cheval – Study of Hamdani, White Arab Stallion – wax
1852	"	Cavalier Hunting a Bull – plaster
1853	"	Horse – bronze
1853	"	Gazelles – bronze
1852	Commissioned by the State	Zebra Attacked by a Panther
1853	Salon	" " " " – plaster
1855	Exposition Universelle	" " " " – bronze
1857	Salon	Bull and Bear – bronze
1857	"	Cow Defending her Calf from a Wolf – plaster
1858	Exhibition of Bordeaux	" " " " " " "

Acquired by the town in that year.

1858 Exhibition of Bordeaux. Bull – bronze. Now in the Bordeaux Museum.

1859	Salon	Mare and her Foal – plaster
1859	"	Cow and Wolf – bronze
1859	"	Dog and Sheep – bronze
1863	"	Jockey on an English Thoroughbred Mare – plaster
1863	Salon	English Thoroughbred Stallion – bronze
1864	"	Children and Dogs – plaster
1865	"	Two Bulls (pair) – plaster. (Bought by the Sultan of Turkey)
1866	Salon	English Thoroughbred Horse – plaster
1866	"	Postilion – bronze group
1866	Commissioned by the Préfecture de la Seine for the central staircase of the Palais de Justice.	Two Lions Seated on their Haunches – stone

(This commission was originally given to Barye in 1859, but the decision was recinded in 1866 when Isidore Bonheur executed the work.)

1867	Salon	Bears and Bull – plaster
1868	"	Dromedary – bronze
1868	"	"Royal" Tiger – plaster
1869	"	Lioness and Cubs – bronze
1869	"	Oxen and Dog – plaster
1870	"	Percheron – plaster
1872	"	Mare and Foal – bronze
1872	"	Cow – bronze
1873	"	"Pépin the Short" in the Arena – plaster

1874	Salon	"Pépin the Short" in the Arena – bronze
1875	"	Head Study – Hound – plaster
1875	"	Head Study – Setter – plaster
1876	"	Cora, 'Setting' Dog – plaster
1876	"	A Lion -- plaster
1877	"	A Family of Tigers – plaster
1878	"	Racehorse – plaster
1879	"	Huntsman (Era of Louis XVth) – bronze
1889	Exposition Universelle	" " " " " "
1879	Salon	A Jockey – bronze
1883	Exposition Nationale	" " "
1889	Exposition Universelle	" " "
1880	Salon	An Arab Hunter – plaster
1881	"	" " " – bronze
1880	"	Horse-dealer – bronze
1881	"	Peasant Leading a Bull – wax
1886	"	A Jockey – bronze
1889	"	Mare – bronze
1889	"	Hounds – bronze
1891	"	Spanish Fighting Bull – bronze
1892	"	African Hunter – silver bronze
1894	"	Stag – bronze
1894	"	Boar – plaster
1895	"	Family of Lions – bronze
1896	"	Sheepdog – plaster
1897	"	"The Game of Polo" – bronze group
1898	"	Horses – bronze
1899	"	Fox on the Alert – marble

WORKS EXHIBITED BY JULES MOIGNIEZ

1855	Exposition Universelle	Pointer and Pheasant — plaster
1859	Salon	〃 〃 〃 — bronze (Purchased by the State for the Salle des Gardes, Compiègne)
1855	Exposition Universelle	Falcon and Weasel Disputing a Skylark — plaster
1859	Salon	Heron — plaster
1861	〃	King Charles Spaniel — bronze
1861	〃	Merino Sheep — wax
1861	〃	Scottish Hound, Setting — wax
1861	〃	Running Dog — wax
1864	〃	Pheasant and Weasel — bronze
1865	〃	Group of Partridges — silver bronze
1866	〃	Egret — silver bronze
1866	〃	Gazelles — wax
1867	〃	Billy and Nanny-goat — wax
1867	〃	Sparrow Fighting — bronze
1876	〃	Bayard, Trotting Horse
1876	〃	Quarrel between Mina and Jaquot, Greyhounds — plaster
1877	Exposition Universelle	Belot, Basset Hound Running — plaster
1877	Salon	Scottish Greyhound and Dog from Havana — plaster
1878	〃	Stallion Chief Baron — plaster
1879	〃	Thio, pure-blooded Newfoundland, and Sultana, Newfoundland Crossed with Mountain Dog — plaster
1880	〃	Tiger Hunt — plaster
1885	〃	Bonhomme, Carthorse Stallion — plaster
1886	〃	Pony and Hound — plaster
1887	〃	Dog and Bitch — plaster
1889	〃	Cupidor, Hound — plaster
1890	〃	Goat with her Kids — plaster
1891	〃	"Mon Chien" — plaster
1892	〃	Roebuck Trotting — plaster

This 19th century excerpt from the following exhibition catalogue of animal painting and sculpture is re-printed by courtesy of the Louvre Museum, Paris:-

**Illustrated Catalogue
of the
Exhibition
of "Animalier" Art
Retrospective**

Organised by the General Association of
Curators of the Public Collections of France

Editions de la
Société des Amis du Muséum

Paris – 1934

19th Century

BARTHOLDI, Frédéric-Auguste, 1834–1904
121 Maquette for the "Lion de Belfort" 1874, Sculpture *(Musée de Belfort)*.

BARYE, Antoine-Louis, 1795–1875
122 Hind, standing. Sculpture. *(Musée du Louvre)*.
123 Bouquetin, standing. Sculpture. *(Musée du Louvre)*.
124 Pointing dog. Sculpture. *(Musée du Louvre)*.
125 Gazelle of Ethiopia. Sculpture. *(Musée du Louvre)*.
126 Jaguar devouring a Hare. Bronze. *(Musée du Louvre)*.
127 Seated Jaguar. Sculpture. *(Musée du Louvre)*.
128 Starving Lion. Sculpture. *(Musée du Louvre)*.
129 Lion and Serpent. Plaster mould. *(Musée du Louvre)*.
129a Black Panther and Spotted Panther. Watercolour. *(Musée du Louvre)*.
130 Tiger devouring an Antelope. Sculpture. *(Musée du Louvre)*.

BELLY, Léon-Adolphe-Auguste, 1827–1877
131 Sleeping Camel. Painting. *(Musée de Tours)*.

BONHEUR, Rosa, 1822–1899
132 Tigers in the Jungle. Painting. *(Musée de Fontainebleau)*.
133 Study of Boars. Painting. *(Musée de Fontainebleau)*.
134 Study of Stags. Painting. *(Musée de Fontainebleau)*.

BOUDIN, Eugène, 1824–1898
135 Game – Still life. Painting. *(Musée de Havre)*.
136 Codfish and Eels – Still life. Painting 1873. *(Musée de Havre)*.

BRASCASSAT, Jacques-Raymond, 1804–1867
137 Study of a Fox. Painting 1837. *(Musée de Nantes)*.
138 Head of a Wolf. Painting 1837. *(Musée de Nantes)*.
139 The Bull. Painting 1842. *(Godard-Desmarest bequest, Musée du Louvre)*.

CAIN, Auguste, 1821–1894
140 Bull for the Trocadero Fountain. Wax mould. *(Musée Carnavalet)*.
141 Lion and Ostrich. Plaster. *(Luxembourg Gardens)*.
142 Lioness carrying her Cub. Plaster. *(Musée Carnavalet)*.

CARPEAUX, Jean-Baptiste, 1827–1875
143 Carriage Horses. Sketch. *(Musée de Valenciennes).*
144 Steer and Horses. Sketch. *(Musée de Valenciennes).*
145 The Duchess of Orleans' Dog. Biscuit porcelain. Mid 19th century. *(Manufacture Nationale de Sèvres).*
146 The Duchess of Orleans' Bitch. Biscuit porcelain. Mid 19th century. *(Manufacture Nationale de Sèvres).*

CLESINGER, Jean-Baptiste, or Auguste, 1814–1883
148 Fight between Roman Bulls. Patinated plaster. Marble shown at the Salon of 1864. *(Musée du Louvre).*

CORDIER, Louis-Henri, 1853–1925
149 3 studies of wild beasts after Assyrian forms in the British Museum. Sketches, wax. *(Musée du Louvre).*
150 Bull. Bronze. *(Musée du Louvre).*

COURBET, Gustave, 1819–1877
151 Hunted Roebuck, listening. Springtime 1867. Salon of 1868. Painting. *(Mme. Vve Boucicaut bequest, Musée du Louvre).*

DECAMPS, Alexandre-Gabriel, 1803–1860
152 Bulldog and Scotch terrier. Painting. *(Anc. Collection S. Goldschmidt. Musée du Louvre).*

DELACROIX, Eugène, 1798–1863
153 Album of animal sketches made at the "Jardin des Plantes". *(Collection D. David-Weill, Neuilly-sur-Seine).*

FREMIET, Emmanuel, 1824–1910
154 Wounded dog. 1851. Bronze. *(Musée du Louvre).*

FROMENTIN, Eugène, 1820–1876
155 Gazelle Hunt in Algeria. 1856. Painting. *(Musée de Nantes)*
156 Hawking in Algeria. Painting. Salon of 1863. *(Collection of Napoleon III. Musée du Louvre).*

GERICAULT, Jean-Louis-André-Théodore, 1791–1824
157 Tigers. Painting. *(Musée de Rouen).*

GIRAUD, Pierre-François-Grégoire, 1783–1836
158 Dog. Marble. Salon of 1827. *(Musée du Louvre).*

HUET, Paul, 1803–1869
162 Study of a Normandy Steer. Painting. *(Donation René-Paul Huet, Musée du Louvre).*

JACQUE, Charles-Emile, 1813–1894
163 Shepherd Leading his Flock of Sheep on the Plains. Painting. Salon of 1861. *(Collection of Napoleon III Musée du Louvre).*

KNIP, née Pauline de COURCELLES, 1781–1851
164 Lyre-birds. Watercolour. *(Musée de Fontainebleau).*

MENE, Pierre Jules, 1810–1879
165 Horses at Liberty. Sculpture. *(Musée due Petit Palais).*
166 End of the Stag-hunt. Sculpture. *(Musée du Petit Palais).*
167 Horses. Sculpture. *(Musée du Petit Palais).*
168 Doe, lying down. Plaster. *(Manufacture Nationale de Sèvres).*
169 Pointer. Plaster

170   Setter. Sculpture. *(Manufacture Nationale de Sèvres)*.

MOREAU, Gustave, 1826–1897

170a   Coiled Snake arising. Design. *(Musée du Louvre)*.

NOEL, Jules-Achille, 1815–1881

170b   Shark. Design. *(Musée du Louvre)*.

170c   Californian Lizard. Design. *(Musée du Louvre)*.

ROLLAND, Auguste, 1797–1859

171   Wild Boars. Painting. *(Donation. The Rolland family. Musée de Metz)*.

BIBLIOGRAPHY

Alexandre, Arséne, *Antoine-Louis Barye* in E. Benezit, Volume 1.
Ballu, Roger, *L'Oeuvre de Barye,* Paris, 1890.
Benezit, E. *Dictionnaire des Peintres, Sculpteurs, Dessinateurs et Graveurs,* Librairie Grund, 1960.
Benge, Glenn Franklyn, *The Sculpture of Antoine-Louis Barye in American Collections,* with a *Catalogue Raisonné,* 1969.
Beiz, Jacques de, *E. Frémiet, un Maitre Imagier,* Paris, 1896.
Bonnat, Léon, in *Gazette des Beaux Arts,* Paris, 1889. P.374-382.
Ciechanowiecki, Andrew S. in *Antiques International* edited by Peter Wilson. p. 308. Michael Joseph.
Champeaux, A. de, *Dictionnaire des Fondeurs, Ciseleurs, Doreurs etc.,* Paris, 1886.
Fauré - Frémiet, Philippe, *Frémiet,* Paris, 1934
Lami, S. *Dictionnaire des Sculpteurs de l'Ecole Française au XIXs.* Volume 1-4, Paris, 1914.
Mills, John W. *The Technique of Casting for Sculpture,* Batsford Reinhold, 1967.
Mirecourt, E. de, in *Les Contemporains* – Rosa Bonheur. p.5-94, Paris, 1856.
Radcliffe, Anthony, *European Bronze Statuettes,* The Connoisseur – Michael Joseph, London, 1966.
Reitlinger, Gerald, *The Economics of Taste,* Barrie and Rockliffe, London.
Saunier, Charles, *Barye,* Rieder et Cie, Paris, 1925.
Savage, George, *A Concise History of Bronzes,* Thames and Hudson, 1968.
Thieme - Becker *Allgemeines Lexikon Der Bildenden Kunstler,* 1933, Leipzig.
Vitry, Paul in *Le Musée d'Art,* Librairie Larousse, Paris.

Catalogues

1) *Barye,* Sculptures, Peintures et Aquarelles des Collections Publiques Françaises 1956/7. Musée du Louvre.
2) *Bronzes d'Art de P.J. Mêne et de Auguste Cain.* Rue de l'Entrepôt, Paris.
3) *The Animaliers, French Animal Sculpture of the 19th Century,* Mallet, Bourdon House, 1962.
4) *Collection de Bronze,* Galerie Georges Petit, 1917.
5) *Les Animaliers,* The Sladmore Gallery, June 1967, April 1968.
6) *Barye to Bugatti,* The Sladmore Gallery, April 1969.
7) *The Horse in Bronze,* The Sladmore Gallery, November 1969.
8) *Catalogue of Sale* 13/2/1970. Sotheby & Co.

```
Book Shelves
NB547 .H67 1971
Horswell, Jane
Bronze sculpture of Les
Animaliers, reference and
price guide
```

DATE DUE